PRIDE IN DEFENCE

PRIDE IN DEFENCE

THE AUSTRALIAN MILITARY & LGBTI SERVICE SINCE 1945

NOAH RISEMAN &
SHIRLEENE ROBINSON

MELBOURNE
UNIVERSITY
PRESS

MELBOURNE UNIVERSITY PRESS
An imprint of Melbourne University Publishing Limited
Level 1, 715 Swanston Street, Carlton, Victoria 3053, Australia
mup-contact@unimelb.edu.au
www.mup.com.au

First published 2020

Text design and typesetting by MK Typesetters
Cover design by Design by Committee
Cover image courtesy bigstockphoto.com
Printed in Australia by McPherson's Printing Group

NATIONAL LIBRARY OF AUSTRALIA
A catalogue record for this book is available from the National Library of Australia

9780522876741 (paperback)
9780522876758 (ebook)

Contents

Acknowledgements vii

Introduction 1

1 Hidden gay and bisexual men, 1944–73 21

2 The women's services, 1950–85 49

3 The heightened ban years, 1974–92 75

4 Challenging the ban 110

5 An era of openness? 1993–2005 138

6 Transgender and intersex service 169

7 A new ADF? 199

Conclusion 231

Bibliography 237

Index 255

Acknowledgements

Pride in Defence is the culmination of years of oral history interviews and archival research that has taken us to all corners of Australia and as far as the United States and Europe. The biggest debt of thanks we owe is to the 140 LGBTI service members who participated in oral history interviews. Some of them had previously talked about their military service many times over. For others, this was the first time that they were opening up about some very dark chapters from their pasts. To all of them, we say thank you for inviting us into your homes and workplaces and sharing a piece of your lives. Thanks also to the other interview participants, be they former politicians, families of ex-service members or former ADF commanders.

The bulk of this research was funded by Australian Research Council Discovery grant DP160103548, and Australian Catholic University provided seed funding for the first two years of the project. This project had the approval of the Departments of Defence and Veterans' Affairs Human Research Ethics Committee (762–14) and Australian Catholic University Human Research Ethics Committee (approval number 2014–22V). The then Vice Chief of the Defence Force, Vice Admiral Ray Griggs, gave command approval to interview currently serving members of the ADF.

We would not have been able to reach so many current and former members of the ADF without the steadfast support of DEFGLIS, and especially its former president, Vince Chong. D'arcy Cartwright was also particularly helpful at networking with a range of currently serving Defence members, and Thomas Hoffman proved a supportive contact within the Defence People Group in the Department of Defence.

Several research assistants contributed to aspects of this project: thanks to Cheryl Ware for coding interviews, to Geraldine Fela for accessing archives in Canberra, Anisa Puri for managing interview data, and most especially to Julien Varrenti-Osmond for spending many weeks going through the LGBTI press and other records at the Australian Lesbian and Gay Archives. Thanks also to the volunteer staff at the Australian Lesbian and Gay Archives for facilitating access to records and for advice, most especially Nick Henderson, Gary Jaynes, Angela Bailey and Kathy Sport. Graham Willett, former president of the Australian Lesbian and Gay Archives, was a partner investigator on the project, and we are grateful for his intellectual contribution and for conducting four of the oral history interviews. In accordance with participants' consent, the majority of oral history interviews have been deposited in the Australian Lesbian and Gay Archives.

We are indebted to colleagues from within our institutions and beyond. From Australian Catholic University, the history team regularly provided enthusiastic support and rigorous peer review. A special thank you to: Shurlee Swain, Melissa Bellanta, Lorinda Cramer, Hannah Forsyth, Meggie Hutchison, Ben Mountford, Nell Musgrove, Maggie Nolan, Jon Piccini and Ellen Warne. Thanks also to Noah's colleagues at other institutions whose intellectual contributions and friendship have made this book possible: Cath Bishop, Fiona Davis, Pat Grimshaw, Ara Keys, Tristan Moss, Patty O'Brien, Yorick Smaal and Ben Wadham.

Shirleene wishes to thank her history friends who span institutions and groups (as well as states, territories and countries): Lisa Featherstone, Lisa Ford, Robert French, Rebecca Jennings, Andy Kaladelfos, Ann McGrath, Scott McKinnon, Clare Monagle, Robert Reynolds, Kay Saunders, Simon Sleight and Yorick Smaal (whose contribution to this field and friendship sees him thanked twice!). The Pride History Group Sydney provides a rich forum to discuss and produce LGBTI history. Shirleene also acknowledges the support of staff and friends at the National Library of Australia who work so tirelessly to collect and preserve Australia's history. Mark Piva's ongoing and steadfast friendship is very much appreciated.

From Melbourne University Publishing, we thank publisher and CEO Nathan Hollier, senior editor Louise Stirling, cover designer Josh Durham and copy-editor Cathryn Game. Thanks also to series editor Joy Damousi, who enthusiastically endorsed this book and encouraged its publication.

Our final thanks go, as always, to the friends and family who have supported us both on the research and writing journey. Noah has befriended so many LGBTI Defence members, past and present, that his Facebook feed is awash with their news and exploits. Noah thanks his family in both Australia and the United States, but most especially his loving partner Michael, who has put up with his trips away, rants about Defence bureaucracy, and exciting stories about the people he met and experiences they shared.

Shirleene thanks the friends she made during the course of this research: Julie Hendy, Christina Dennis and Carole Popham. Getting to know you has been one of the greatest joys of this project—your perspectives and generosity are much appreciated. She also thanks Alex Greenwich, whose commitment to equality is inspiring. She gratefully acknowledges family and friends who have supported her research with interest and patience: Kay Harrison and Sophie Partridge, Paris Lehn, Ky Menzies, Morgan Spearritt, Leah Newman, Helen Ross-Browne, Sarah Taylor, Julie Ustinoff and Emily Wilson—thank you. Sarah Midgley, as always, has been central to this project. Her love and kindness make all things possible.

The text of a few chapters has derived in part from the following previous publications:

Riseman, Noah. 'Activism and Australia's ban on gay, lesbian and bisexual military service in the 1970s–80s.' *Australian Feminist Studies* 33, no. 95 (2018): 147–63. Copyright © Taylor & Francis. http://www.tandfonline.com/10.1080/08164649.2018.1498728

—— 'Hunting gays and lesbians in the Australian Defence Force, 1974–1992.' *Journal of the History of Sexuality* 28, no. 3 (2019): 325–56. Copyright © 2019 University of Texas Press. All rights reserved. Used by permission.

—— '"Just another start to the denigration of Anzac Day": Evolving commemorations of LGBTI military service.' *Australian Historical Studies* 48, no. 1 (2017): 35–51. Copyright © Taylor & Francis. http://www.tandfonline.com/10.1080/1031461X.2016.1251476

—— 'Outmanoeuvring defence: The Australian debates over gay and lesbian military service, 1992.' *Australian Journal of Politics and History* 61, no. 4 (2015): 562–75. Copyright © John Wiley & Sons. https://onlinelibrary.wiley.com/doi/10.1111/ajph.12119

—— 'Transgender policy in the Australian Defence Force: Medicalization and its discontents.' *International Journal of Transgenderism* 17, no. 3–4 (2016): 141–54. Copyright © Taylor & Francis. http://www.tandfonline.com/10.1080/15532739.2016.1227759

Robinson, Shirleene. 'Sexuality at a cost: Lesbian servicewomen in the Australian military, 1960s–1980s', in *Beyond Combat: Australian Military Activity Away from the Battlefield*, ed. Tristan Moss and Tom Richardson (Sydney: NewSouth Publishing, 2018), pp. 45–58

Introduction

In May 1982, the *Weekend Australian* published a letter to the editor written by a veteran named Bob Herbert. Although heterosexual himself, he alluded to a lengthy history of gay military service, asserting that he had encountered numerous homosexual men during World War II who were 'generally tolerated so long as they took no for an answer'. Herbert continued, however, to reveal the stark limitations of this tolerance by describing what happened to an officer who was believed to have engaged in same-sex activity and was court martialled for buggery. He wrote:

> We rookies were all ceremonially paraded while the poor man was marched on under armed escort and made to stand to attention, hatless, while the court-martial finding and sentence was read out. A drum was then rolled and the CO commenced stripping the epaulets of rank and various badges from the culprit's uniform. This humiliation was never completed; the victim fainted and was carried from the parade ground.[1]

Flash forward to 2016, and life for gay men as well as lesbian, bisexual, transgender and intersex members of the Australian Defence Force (ADF) was very different. That year, openly gay Warrant Officer (now Lieutenant Commander) Stuart O'Brien of the Royal Australian Navy (RAN) marched into a different public ceremony. At Government House in Sydney, the Deputy Governor of New South Wales pinned a medal of the Order of Australia (OAM) in the Military Division on to O'Brien's uniform. The citation for the honour read:

> As founder and chairperson of the Defence Gay and Lesbian Information Service, Warrant Officer O'Brien has equally

> enhanced the lives and careers of these members and their families and helped foster a culture of inclusion and respect within Defence. His professionalism and leadership are in the finest traditions of the Royal Australian Navy and the Australian Defence Force.[2]

These two public rituals and the narratives underpinning them capture a long and often unacknowledged history of lesbian, gay, bisexual, transgender and intersex (LGBTI) service in the ADF.[3] They also illustrate the way that the ADF has changed its policies towards LGBTI people. The ADF has adopted three broad approaches to LGBTI service in the twentieth and twenty-first centuries: rejection, tolerance and inclusion. These have not been mutually exclusive or practised uniformly over time. There are numerous examples of officials tolerating some LGBTI members while discharging others. Even as the ADF permitted LGB members to serve openly after November 1992, rules it adopted in April 2000 excluded transgender service members until September 2010. Yet the two starkly different ceremonies highlighted above also demonstrate the dramatic culture change in the ADF since World War II. An institution that once ostracised and expelled LGBTI members now nominates them for prestigious Australian honours. Just as this transformation has been both difficult and uneven, it has also come at great personal, professional and financial cost for LGBTI members who were targeted, persecuted and compelled to discharge because of their sexuality or gender identity. More often than not, it was not the benevolence of the ADF or commanding officers that drove change. Rather, it was concerted efforts by LGBTI service members and the intervention of external bodies—especially the Australian Human Rights Commission—that forced the ADF and Commonwealth Government to confront archaic policies that discriminated against LGBTI service members.

This book analyses the history of LGBTI military service in the ADF from the end of World War II until the present. We recognise that the term 'LGBTI' is historically contingent and that sexuality and gender identity have been shifting concepts whose meaning has evolved over time, including during the period considered in the book. As Ken Plummer has asserted about homosexuality, 'Throughout time and space the pleasures and displeasures of erotic experience between the same genders have certainly existed; but in every culture such experiences both create and respond to a wider set of cultural meanings.'[4] This concept should also be expanded to apply to gender identity. While there have always been individuals whose sex assigned at birth has not matched their actual gender identity, and individuals who have reacted to this in a variety of ways, the concept of being transgender is a relatively modern one. The ADF adopted different policies and practices towards individuals on the basis of sexuality, gender and intersex status throughout the period covered in this book. To account for this, we also use the nomenclature 'LGB', 'transgender', 'non-binary' and 'intersex' at times to mark these variations in treatment clearly.

Pride in Defence explores all dimensions of LGBTI service, both inside and outside the ADF. From inside the ADF, this entails examining the evolving policies, practices and lived experiences of LGBTI Defence members. The book explores the contributions LGBTI members have made to all three branches of service. It considers LGBTI subcultures, including distinctions across the services and the different experiences of men and women. The book also explores how the politics of LGBTI service evolved outside the ADF, such as the activism that challenged the LGB service ban and the political debates that were waged over LGBTI inclusion. Examining the intersection between the ADF and the political realm reveals much about the ways in which Australians have imagined the ADF as being reflective (or not) of Australian society.

In recent years the ADF leadership has been vocal about supporting LGBTI inclusion, concurrently acknowledging the institution's troubled history and the progress made. In 2015, Army Regimental Sergeant Major (RSM) David Ashley summarised it well: 'Society's changed and so has the Australian Defence Force ... I'm ashamed of some of those things that put a dark spot on Defence and on Army. I can tell you this, though: we have changed. And if we find this abuse, and we have proven this with our recent actions, we will take decisive action.'[5] Ashley is correct that there have been substantial changes within the ADF, but to date there has been little scholarly work studying exactly *what* those changes have been, *why* they were implemented, and *how* they have affected Defence personnel. Drawing on a mix of oral history interviews with 140 LGBTI current and ex-service members, media reports from the mainstream and LGBTI press, Defence policy documents, personal archives and other records from the National Archives of Australia and Australian Lesbian and Gay Archives, this book showcases the complex, nuanced and constantly shifting dynamics of LGBTI service in the ADF. It also reveals that while the ADF has a largely positive story to tell about culture change, there are still steps to be taken in relation to transgender, intersex and gender-diverse service.

Literature review

Internationally, some of the earliest books on gay and lesbian history focused on military service. There were sound reasons for this: service during World

War II created a climate where many men and women were able to discover others who shared their desires. The first texts on World War II and post-war military service all drew on oral histories and policy documents to explore the experiences of gays and lesbians in the US forces. They all explored multiple themes about LGBTI military service: the influence of psychological discourse about homosexuality; different treatment and anxieties about women's sexuality versus gay men; discretion versus flamboyance; military police procedures; debates among policy-makers over how to grapple with the 'problem' of homosexuality in defence forces; and LGB service members' challenges hiding their sexuality.[6] Later studies about homosexuality in the Canadian and British forces during World Wars I and II drew on a similar mix of oral histories, newspaper reports, government documents and military records, and pursued similar thematic lines of inquiry.[7]

Marking a potential historiographic turn, Kellie Wilson-Buford's recent book *Policing Sex and Marriage in the American Military* examined homosexuality as one of several 'moral' issues that US military law regulated in the forces from 1950 to 2000. She showed how the application of military laws against homosexuality, adultery, overseas brides and sexual assault have consistently reinforced and policed 'traditional' values about sex, sexuality, gender and the family.[8] American and British LGB veterans have also been the subject of several biographies and autobiographies. These books aimed for readers to understand the challenges and personal costs borne by LGB service members, who risked discharge if their sexuality were discovered.[9]

Transgender military service has only recently become a topic of public interest and, due to a climate of transphobic prejudice that still exists, it has only been very recently that many transgender people have been comfortable to share their stories. Consequently, globally there are fewer transgender service members' (auto)biographies. From the United States, the only known biography from a transgender service person is Kristin Beck's story of training as a Navy SEAL in 1991 to her decision to discharge in 2011 in order to undergo a process of gender transition. Beck's struggle with her gender identity was intricately linked with her military service, as she chose a hypermasculine, dangerous role to try to 'prove' her masculinity (unsuccessfully).[10] Caroline Paige, the first openly transgender officer in the British Armed Forces, narrates a similar life struggle with her gender identity, culminating in her transition in 1999. Her book also focuses on her valiant work as a pilot in the Gulf War and the Balkans and her service post-transition as a helicopter pilot in Iraq and Afghanistan.[11] Chapter 6 of this book shows that the struggles and service motivations presented by both Beck and Paige have echoes among Australian transgender personnel. In fact, Paige's public outing was likely the reason Australia implemented an explicit transgender ban in April 2000.

In Australia, too, World War II was the first site of scholarly interest for historians of LGBTI military service. Independent queer historians were at the forefront of recording anecdotes and publishing short pieces in the LGBTI press. Probably the first published story of gay service in World War II was a first-hand account written by 'Hadrian' and published in the Western Australian *Campaign Circular* in November–December 1972. Hadrian recalled an Army toilet in Atherton, Queensland, that served as a beat—a public place frequented by gay and bisexual men in search of sex. He also remarked, 'I would say that the incidence of homosexuality in the forces is greater than the conventional 5%. The military provides an all-male environment and as long as the soldier is reasonably careful and discreet, problems very seldom arise.'[12]

It was not until the 1990s that gay and lesbian history emerged as a serious academic pursuit, and academic historians began to research histories of LGB people in the ADF. Ruth Ford was the first, focusing on the experiences of lesbians in the World War II women's services. She showed that this single-sex environment was one where many women—sometimes unexpectedly—found themselves attracted to each other and were able to explore their sexuality. The services did not have formal policies against homosexual women. There were, however, always prevalent underlying anxieties about the masculinisation of women in the services or the corrupting potential of lesbianism. Commanding officers regularly took action against suspected lesbians, which could include discharging them.[13] Ford was also the first scholar to look at the post-war women's services, using oral histories to uncover the ways in which lesbians secretly forged relationships in the 1950s and 1960s, even while authorities targeted them more aggressively.[14] Chapter 2 of this book extends Ford's work, using new oral histories and declassified documents to analyse the so-called witch-hunts and ways in which women discovered and explored their sexualities in stealth in the Women's Royal Australian Naval Service (WRANS), Women's Royal Australian Air Force (WRAAF) and Women's Royal Australian Army Corps (WRAAC).

Yorick Smaal was the first historian to publish a monograph on homosexuality in the Australian Army. *Sex, Soldiers and the South Pacific, 1939–45* (2015) focuses on the policing of male homosexuality in Brisbane and Papua New Guinea during World War II. Drawing on discipline files, Brisbane police and court records and psychiatric literature, Smaal showed the ways in which soldiers discreetly sought out and experienced homosexual encounters both while on leave in Brisbane and on the front lines. He also revealed that commanding officers often tolerated known homosexuals, and rituals like drag shows were public performances and safe spaces for men to subvert gender norms. It was only when the Americans complained about the presence of homosexual men in Papua New Guinea that the Australian Army contemplated the homosexual 'problem' and formally adopted a policy explicitly targeting gay men.[15]

As in the United States, there are a small number of Australian autobiographical accounts of LGBTI service members, but they differ in scope. In American LGBTI memoirs, military service is usually the central topic of the narrative. In Australian accounts, military service tends to be one chapter or one aspect of a broader life narrative about an individual coming to terms with their sexuality, gender identity or intersex variation. Only Roderic Anderson has written about World War II, having served in the RAAF in Australia and present-day Indonesia. Anderson encountered homoerotic behaviour at many of his postings, and he had his first homosexual experiences during the war.[16] Gay Aboriginal man Noel Tovey describes his six months of National Service in the RAAF in 1953 as a liberating experience, but does not comment on homosexuality during his service.[17] Autobiographies of gay Vietnam veterans describe the war as a pivotal moment in their lives but as a generally sexless affair.[18] One notable exception is Lorenzo Montesini, who met his partner on a beach in Vung Tau in 1967 before they spent a steamy night together. That was the beginning of a tumultuous relationship of over twenty-seven years, ending when his partner died of an AIDS-related illness.[19]

There are also autobiographies from Australian transgender and intersex ex-service members. The most famous is the award-winning *Katherine's Diary* (1992), which narrates Katherine Cummings's life story from being raised in Kiribati, migrating to Australia as a child during World War II, training as a librarian, living overseas and eventually transitioning gender in 1987 at the age of fifty-two. Briefly mentioned in one chapter is Katherine's time as a national serviceman in the Navy in 1954.[20] The other, more intriguing story is Peter Stirling's book *So Different*, marketed as the autobiography of a 'transsexual' man. It was published in 1989 when society emphasised very different understandings about sex, gender, gender identity and the body. Stirling was raised as a female and served in the WRAAF from 1954 to 1955. Stirling fell in love with another servicewoman, the two even going absent without leave (AWL) together. Stirling was later charged because of the relationship and discharged. What might otherwise be seen as a lesbian relationship in the WRAAF is more complicated. Ten years later, Stirling was diagnosed as having XXY chromosomes and had medical interventions to transition to male.[21] Having an XXY chromosomal pattern is now more widely known as Klinefelter syndrome and is recognised as an intersex variation. Stirling's remarkable autobiography provides insights into the experiences of both lesbians and intersex military service members.

Pride in Defence is the culmination of the first research project specifically investigating the history of Australian LGBTI service in the post-World War II era. The project began in 2014 and was funded from 2016 to 2019 by an Australian Research Council Discovery grant. We have already produced a series of articles and book chapters addressing specific aspects of that history, including: the politics and debates over lifting Australia's LGB ban in 1992; policy changes in relation to transgender service; the experiences of lesbian servicewomen in the 1970s; commemorations of LGBTI service on Anzac Day and at Sydney Gay and Lesbian Mardi Gras; and activism against the LGB ban in the 1970s.[22] In 2018 we also published *Serving in Silence? Australian LGBT Servicemen and Women*, presenting the life stories of fourteen LGBT servicemen and women. The biographical approach showcased how military service affected LGBT service members' lives, the challenges they had to overcome, valiant service they performed and relationships forged.[23] Here, we extend this scholarship and bring together new oral histories and hitherto unexamined documents to produce the first comprehensive history of LGBTI military service in post-war Australia.

Military sociology

Framing much of this research is literature on militaries, citizenship, gender and sexuality. Military sociologists have regularly debated the relationship between armed forces and civilian society, particularly considering the question of whether militaries should be seen as reflections of the nation-state or must necessarily be conceptualised, constructed and regulated differently. Even Carl von Clausewitz argues that the nature and organisation of the state has a major influence on the way an army functions and wages war.[24] Among modern military sociologists, Morris Janowitz leads the field with his analysis of the relationship between citizenship and military service. He argues that since the American and French revolutions, military service in the West has been constructed as a duty of (male) citizenship with attendant obligations and veteran privileges.[25] It is for this reason that marginalised groups around the world—especially racial minorities—have regularly used their military service to argue for equal citizenship rights, with varying degrees of success.[26] In Australian history this was most prominent in the case of Aboriginal and Torres Strait Islander people, who used their military service as one argument for citizenship rights after World Wars I and II.[27]

The soldier-citizen framework applies to LGBTI people as well, but in a different way. Traditionally, racial minorities have made the soldier-citizen argument like this: because they served in defence forces, and often were subject to conscription, they should be granted equal citizenship rights in civilian society. For LGBTI people around the world, the argument has gone in reverse: because they are citizens, they should be allowed to serve openly in the armed forces. In Australia, only occasionally since 1992 have LGBTI service members argued that because they can serve they should be afforded equal rights. This was most prominent in the fight for same-sex couple recognition in the ADF, and it was a somewhat muted argument used in the push for marriage equality.[28]

Australia's culture of veteran entitlement and the status afforded to military service has derived substantially from the Anzac mythology.[29] The term 'ANZAC' refers to the Australian and New Zealand Army Corps, which landed at Gallipoli during World War I on 25 April 1915. As news filtered back to Australia (and New Zealand) about the prowess of this force, a mythology grew that persists to this day, defining a particular iconic Australian 'digger' as a soldier who, by extension, is the embodiment of Australian nationhood.[30] The Anzac mythology depicts a particular, exclusive image of the ideal Australian serviceman as 'the stereotypical representation of the ideal Australian as a tall, tough, laconic, hard-drinking, hard-swearing, hard-gambling, independent, resourceful, anti-authoritarian, manual labouring, itinerant, white male'.[31] To this list could be added heterosexual and cisgender (a person whose gender identity aligns with the sex assigned at birth). The strength of the Anzac mythology has waxed and waned over the past century,[32] but its presence has been constant and it has strongly influenced constructs of Australia's national identity and perceptions of the ADF and its members. Indeed, James Brown, a retired Army officer and past president of the Returned and Services League (RSL) NSW, argues that members of the ADF today struggle in part because they still benchmark themselves against this (false) mythology.[33]

Although there is something distinctly Australian about the Anzac mythology, it is a derivation of global military cultures, which are highly gendered. Scholars such as Joshua Goldstein, Cynthia Enloe, Teemu Tallberg and Johanna Valenius have documented the masculine ideologies underpinning militaries.[34] Kellie Wilson-Buford summarises: 'Militaries enact rites of passage that force men to prove their masculinity by renouncing weakness, sadness, feminine traits and characteristics, and other qualities that typically were viewed as feminine.'[35] Australian sociologists such as Jyonah Jericho, Katerina Agostino and Ben Wadham similarly argue that the ADF has always been a hegemonic masculine institution, with power structures and traditions favouring martial masculinity over traits associated with femininity.[36] Stereotypes of gay men as camp, weak and feminine have meant that they have historically been perceived as incompatible with military service. Wadham describes the ADF as a homosocial fratriarchy, where 'fraternity is crucial to strong teamwork but it can also culminate in very strong them and us attitudes, often inferiorising or denigrating the other'.[37] Wadham argues that the fratriarchal culture has tolerated or even promoted racism and sexism because any difference constitutes a challenge. Homosexuality and diverse gender identities represent other major sites of difference, and they could also be portrayed as a threat to the bonding and the homosociality associated with military culture.

In the post-World War II era, as various social movements around the world forced Western societies to rethink attitudes and policies towards race, gender and sexuality, militaries have also had to grapple with the forces of change. This process accelerated especially after the Cold War, when the easing threat of total war and changing nature of military operations meant that armed forces came under increasing pressure to reform.[38] Political scientist Hugh Smith argued in 1995 that 'as the memory of past war and the threat of future war have receded, the ADF has become more open to change and more susceptible to influences from Australian society'.[39] Among the changes to the ADF in the 1980s and early 1990s were a shift to seeing military service more as an occupation/profession; the gradual disbandment of the separate women's services, completed in 1985; and eventually the lifting of the ban on LGB service in November 1992. These reforms have always been a process necessitating accompanying culture change, with numerous factors helping or hindering the process.

Most important has been the way service members have *perceived* these reforms: either as positive steps to increase the ADF's capabilities, or as impositions from government or a disconnected top brass. Ben Wadham et al.'s study on the ADF's integration of women from the 1970s through the present reveals the continuing embedded masculine culture within the ADF. Rather than challenge the ADF's masculine ideology, women's service is constructed in binary opposition: 'women's rights versus capacity, integration versus transformation, and the functional versus societal dialectic'.[40] Although ADF policies have, over time, become more gender neutral, Wadham et al. argue that neutrality has really required women to adapt to the institution's masculine culture rather than representing any significant culture change.

Although Wadham et al.'s assessment of women's service suggests that the ADF is slow at (or even incapable of) culture change, this book shows the LGBTI experience to be somewhat different. As chapters 4 and 5 show, there is ample evidence that in 1992 the vast majority of Defence members opposed lifting the ban on LGB service. Yet the oral histories of LGB service members from that era reveal more nuance, as many found that their immediate co-workers were accepting of their sexuality. Even as far back as World War II, there are reports of service members expressing no difficulties with some individuals' homosexuality. Over time, as more courageous LGBTI service members have been open about their sexuality or gender identity, there have been fewer reported cases of discrimination. Indeed, as more LGBTI people have served openly, they have challenged the stereotypes and binaries associated with homosexuality, masculinity, femininity and gender more broadly. Homophobia and transphobia are still present and perhaps always will be, but oral histories of current LGBTI members suggest that prejudice and discrimination have been on the wane for at least the last fifteen years.

Oral history theory and methodology

Oral history has proven especially popular among LGBTI historians because, for so long, there have been so many silences in written records. The mass digitisation of newspapers and other documents has opened new possibilities for uncovering histories of sexuality and gender non-normativity, yet until recently the most common archival sites for records on gay men were either in police files or records on psychology. Female homosexuality was never a crime in Australia, so the historical silences surrounding lesbians are more pronounced.[41] Transgender and intersex people have been even more obscured in history, and it is only now that historians are beginning to comb records to find examples of gender non-normativity to construct trans-historicities.[42]

Searching for LGBTI histories in the ADF, one encounters similar archival barriers. The National Archives of Australia holds ADF policy documents on homosexuality, although these are primarily from the period 1974–92. Documents for the post-1992 era are harder to come by because the *Archives Act 1983* (Cwlth) restricts the open access period to the pre-2000 period, although Freedom of Information requests to the Department of Defence have been helpful in uncovering more recent documents. The most obvious place to look for historical records relating to homosexuality or transgender people are the respective military police files covering investigations into homosexuality pre-1992. Owing to legitimate privacy provisions in the *Archives Act*, these records are generally inaccessible to researchers, and the few we could access are heavily redacted. We were, however, able to source some valuable records from the personal archives kept by LGBTI ex-service members.[43] The ADF records we accessed were most useful to reconstruct and analyse the institutional responses to homosexuality and transgender people.

Our other principal source was oral history interviews, which were vital in recovering the personal experiences, perspectives, motivations and life journeys of LGBTI service members. Oral histories also exposed anecdotes about LGBTI subcultures and military police practices—information that was not contained in police reports. A considerable body of international research, including work by Madeline Davis and Elizabeth Lapovsky Kennedy, has shown the restorative potential of oral history as a method. By allowing marginalised communities and individuals to frame their historical experiences, official silences are broken and a fuller, more personal understanding emerges.[44] As Lapovsky Kennedy has observed, oral history addresses questions of 'how individuals cope with and resist heterosexism and homophobia. How do individuals decide to construct and express their identities?'[45] Oral history interviews for this book—and the broader project from which it emerged—were framed by the life story method, allowing narrators to locate military service within the context and contours of a wider life. What interviewees did and did not opt to reveal and focus on were all instructive.

We conducted oral history interviews with eighty-nine former and fifty-one current LGBTI Defence members, including reservists. The interview protocol received approval from the Australian Departments of Defence and Veterans' Affairs Human Research Ethics Committee, along with the sponsorship of the Department of Defence Diversity Directorate and the command approval of the Vice Chief of the Defence Force. The breakdown of interviews is shown in table 1.

Table 1: Breakdown of oral history interviews conducted with LGBTI Defence members*

Category/service	**Navy**	**Army**	**RAAF**	**Total**
Gay men	20	29	37	86
Lesbian or gay women	7	18	12	37
Bisexual men	1	1	1	3
Bisexual women		1		1
Trans men		1	3	4
Trans women	1	2	3	6
Trans non-binary		1	1	2
Intersex			1	1
Total LGBTI	**29**	**53**	**58**	**140**

*In cases where a member served in more than one service, the one recorded is the more recent.

We also interviewed three family or close friends of deceased LGBTI ex-service members and a small number of straight ex-service members who discussed policies and practices related to LGBTI service. We contacted the interview participants through a variety of means: word of mouth with other current or past LGBTI Defence members; advertisements in the LGBTI press; and for current members especially, through the Defence LGBTI Information Service (DEFGLIS). All interview participants had the option of being identified or using a pseudonym; those with pseudonyms are denoted by quotation marks. We use the oral histories in dialogue with the archival sources and with each other to reveal common themes and reconstruct a dominant narrative of the past while also acknowledging the diversity of experiences within the dominant narrative. We found ourselves in agreement with queer historians Horacio N. Roque Ramirez and Nan Alamilla Boyd, who found that the 'liberating quality of many queer narrators' stories reveals the intensity and drama of the oral history exchange—and the bond often formed between narrator and researcher'.[46]

One of the perpetual challenges oral historians grapple with is the reliability of memory, and theorists have written extensively about this complication.[47] It is not the purpose of this section to provide an in-depth theoretical or methodological exploration of the reliability of oral histories. Even so, given that the book relies heavily on oral history interviews, there is the need for some discussion about composure, memory and how these two interlinked concepts relate to oral histories of LGBTI military service.

There is an inherent tension or even paradox at play: when researchers have only a small number of oral histories, they face accusations that they might not be a representative sample. When there are many oral histories that corroborate common tropes, the interviews might be accused of constituting composed memories and therefore be inaccurate and unreliable. Composure came to the fore of oral history theory in the 1990s and has a double meaning. First it refers to how a narrator constructs, or composes, a narrative about themselves; second, it describes how narrators seek a sense of poise (composure) as they tell the story.[48] Much of the literature on composure focuses on two aspects. The first is how narrators tend to compose their memories around what is publicly acceptable or, alternatively, how narrators will seek out public audiences that affirm their identities and memories.[49] As Penny Summerfield notes, 'If they [narrators] cannot draw on an appropriate public account, their response is to seek to justify their deviation, or to press their memories into alternative frameworks, or to be able to express their stories only in fragmentary and deflected accounts.'[50] Indeed, as Alamilla Boyd observes in relation to queer oral histories, many narrators try to compose their understandings of self around their LGBTI identities, potentially (re)shaping the way they discuss life events.[51]

The other aspect of composure that has drawn significant critical attention, and which is more relevant to this book, is the relationship between dominant narratives and memories. There does not have to be a public narrative for people to compose their memories. Rather, oral historians argue that when there *is* a dominant public narrative, it may influence composure. Alistair Thomson's research with Australian World War I veterans in the 1980s found there was a strong link between dominant public narratives of the war and Anzac mythology, and the way that veterans composed their own memories. Thomson realised the potency of pop culture and public discourse to shape the way individuals compose their memories to align with dominant narratives.[52] Thomson drew heavily on the work of the United Kingdom's Popular Memory Group, which argued in 1982 that 'Private memories cannot, in concrete studies, be readily unscrambled from the effects of dominant historical discourses. It is often these that supply the very terms by which a private history is thought through.'[53] Other scholars such as Wolf Kansteiner, Anna Green and Penny Summerfield have theorised the relationships between individual and collective memories, and how public discourse influences and shapes composure.[54] They conclude that individual and collective memories function in a dialogic relationship. As Lynn Abrams effectively summarises, 'People do not merely absorb dominant discourses, use them to shape their own life narratives and spout them back at the interviewer. Clearly there are gaps and tensions between individual accounts and dominant or public representations which may emerge in the interview context. These may be difficult to traverse.'[55]

Of course, composure is often overplayed as an explanation for how and why individuals remember particular events. Anna Green asserts the importance of affirming interview participants' agency and 'the capacity of the conscious self to contest and critique cultural scripts or discourses'. Indeed, Green notes that often there is tension between individual and collective memories, with individuals *aiming* to disrupt the dominant accepted narratives.[56] With regard to the concept of composure and LGBTI Defence members, when narrators reported similar experiences (e.g. in relation to police investigations), there are essentially two possible reasons. The first is to accept that through decades of telling and retelling stories, LGBTI ex-service members have composed their memories around dominant narratives of witch-hunts, police interrogations and devastating separations from the ADF. The second and most likely possibility, given the limited visibility of LGBTI service personnel in wider public accounts, is that ADF policies, practices and culture were so engrained that it should be expected that LGBTI service members witnessed similar events.

This book accepts the second possibility for a few reasons while still acknowledging that composure could play a role in shaping some LGB ex-service members' memories. As noted, there has not been a dominant narrative of Australian LGBTI military service until the intervention of this research project. Our interview participants generally had clear, structured narratives of their lives and military service. Wolf Kansteiner notes that small groups with shared traumas can influence collective or national memory only if they have the means to share and disseminate those memories. Moreover, there needs to be a contemporary interest in those histories for such groups to gain traction.[57] On the one hand, there is interest in contemporary Australia to redress historic wrongs about homosexuality; on the other hand, until this project there was no interest in the historic persecution of LGBTI people in the ADF.[58] Essentially, the histories of LGBTI service were not being shared or recorded, so there was little scope for public discourse to shape ex-service members' composure. If anything, the news, public discourse and pop cultural references from the United States about the 'don't ask, don't tell' policy has been more prolific in Australia than any discussion of its own LGBTI Defence history.

The final reason to accept the veracity of the oral histories is that, as this book shows, they generally align with the written records. Accepting the oral histories as valid sources means that, as Alamilla Boyd advocates, the narrators 'could verify the accuracy of the "data to be offered up for the historical record". In this way, the narrators themselves provided a "reality check".'[59] The oral histories of course have variations that account for the individual circumstances of different LGBTI members. Yet, following the advice of oral historian Trevor Lummis, there are enough patterns to make generalisations on the basis of this representative sample. Indeed, Lummis argues that common trends should not necessarily be used to discount the testimonies, but rather, 'If a sample is conformable to known trends, one can have some confidence that the internal distinctions will reflect real distinctions.'[60]

Structure

Pride in Defence proceeds chronologically and thematically, with each chapter focusing on major issues that confronted the ADF and/or LGBTI Defence members during particular periods of service. Chapter 1 begins with World War II in Papua New Guinea and explains the introduction of the Australian Army's first explicit policy on gay military service. It then shifts to examine gay and bisexual servicemen from the period until the early 1970s. Military documents are mostly silent, but there are occasional records such as a series of investigations into homosexual behaviour, naval courts martial for acts of gross indecency or indecent assaults, and occasional newspaper reports of soldiers arrested for visiting beats. The silences themselves are also telling, for the military's general inattention to homosexuality meant that there were often opportunities for gay and bisexual men to have sexual encounters both within and outside the services.

Although policies specifically targeted male homosexuality, from early on anxieties about women's sexuality led officials in the WRANS, WRAAC and WRAAF to embark on witch-hunts to investigate and expel women who were suspected of same-sex activity. Chapter 2 explores the experiences of lesbian women in the women's services, which proved to be sites where a vibrant lesbian subculture thrived. The women's services also provided women with a way of transcending some of the gendered expectations and limitations existent in Australia between the 1950s and 1980s. Yet being identified as someone who had engaged in same-sex activity had serious consequences for such women. In many ways, the treatment meted out to lesbian and bisexual women in the decades after World War II formed the template that was widened to persecute gay and bisexual men after 1974.

Chapter 3 focuses on the policies, practices and life for LGB members from 1974 to 1992, when authorities most strongly enforced the ban on their service. The chapter pays particular attention to the policy instruments and their interpretations during the era of the ban, including what the rules permitted in investigations versus how investigations usually proceeded. The examples from oral histories show how the ban operated in practice, including how gay and bisexual men and women forged secret lives while wary of the ever-watchful eye of the military police.

The LGB ban came under sporadic challenge from gay and lesbian activists in the 1970s, from LGB service members in the 1980s, and from the political class in the early 1990s. Chapter 4 examines all of these pushes for change and the process that began with challenges by two dismissed lesbians in the Human Rights and Equal Opportunity Commission in 1991, which culminated in the Keating government's decision to lift the LGB ban on 23 November 1992.

Although the ban might have been lifted, during the period from 1993 to 2005 the ADF was still not a welcoming space for LGBTI personnel. Chapter 5 explores the ADF policies and experiences of LGB Defence members during this era. It shows examples of those men and women who remained closeted for fear of persecution and those pioneers brave enough to come out and challenge the discriminatory policies around recognition of same-sex couples.

Until the 1990s there was little mention of transgender or intersex service, but there is a much longer, silent history of transgender people in the ADF. Chapter 6 draws on stories of transgender members both before the transgender ban was lifted—pieced together from oral histories and media reports—as well as the changing experiences of transgender members since 2010. The chapter also pays attention to the even more unknown history of intersex members. Using oral history and media reports, it shows how intersex variations were seen as 'medical conditions' and therefore treated differently from LGBT people.

Chapter 7 examines the period since 2005, showing the continuing varying experiences of LGBTI members together with a general trend towards more acceptance. This is manifested most profoundly through the leadership of advocacy organisation DEFGLIS and through symbolic gestures such as permitting ADF members to march in the Sydney Gay and Lesbian Mardi Gras in uniform. Interviews with some of the current and recent cadets at the Australian Defence Force Academy show areas where there is continuing work to be done, particularly to support transgender, non-binary and gender-diverse members.

The conclusion makes final reflections on the overall transformation of the ADF in the past seventy-five years, especially in its relationship to LGBTI personnel but also more generally. It links the ADF to international examples of LGBTI military service and to global trends in LGBTI affairs. At times Australia has led the world, and there is much to be gleaned from appreciating the ongoing evolution of the ADF.

Although this book is a comprehensive history of LGBTI service in the ADF, we acknowledge that it is not and never can be an entire history. There are as many stories as there are LGBTI Defence members. LGBTI readers might find resonance with much of what is in this book while others might have other stories that further complement and supplement what we tell. We anticipate, though, that this book will lay the groundwork for members of the ADF to understand this part of its history and to take lessons from that history for the future. We hope also that Australians more broadly will learn about a long history of unacknowledged service and broaden their understanding about the ADF and LGBTI Australians.

Notes

1 Bob Herbert, 'Still shocked', *Weekend Australian* (1–2 May 1982), p. 11.

2 Vince Chong, 'DEFGLIS founder receives Order of Australia honours', DEFGLIS, https://www.defglis.com.au/index.php/news/233-defglis-founder-receives-order-of-australia-honours, retrieved 5 November 2018.

3 The ADF was not formally constituted until 1976, but we use the term for the sake of convenience when referring to the long tri-service history of the Royal Australian Navy, Australian Army and Royal Australian Air Force (RAAF). When specifically discussing the pre-1976 era, we use the more appropriate terminology applicable to the relevant service.

4 Ken Plummer, *Telling Sexual Stories: Power, Change and Social Worlds* (London and New York: Routledge, 1995), p. 81.

5 'Senior Defence Force members to lead ADF contingent in Mardi Gras Parade', *RN Breakfast*, 6 March 2015, http://www.abc.net.au/radionational/programs/breakfast/australian-defence-force-marches-in/6285110, retrieved 2 July 2015.

6 Allan Bérubé, *Coming out under Fire: The History of Gay Men and Women in World War Two* (New York: Free Press, 1990); Randy Shilts, *Conduct Unbecoming: Lesbians and Gays in the US Military, Vietnam to the Persian Gulf* (London: Penguin Books, 1993); Nathaniel Frank, *Unfriendly Fire: How the Gay Ban Undermines the Military and Weakens America* (New York: Thomas Dunne Books, 2009); James Parco and David Levy (eds), *Evolution of Government Policy Towards Homosexuality in the US Military: The Rise and Fall of DADT* (Abingdon, UK, and New York: Routledge, 2014); Mary Ann Humphrey, *My Country, My Right to Serve: Experiences of Gay Men and Women in*

the Military, World War II to the Present (New York: HarperCollins, 1990); Steve Estes, *Ask and Tell: Gay and Lesbian Veterans Speak Out* (Chapel Hill, NC: University of North Carolina Press, 2007); Steven Zeeland, *Sailors and Sexual Identity: Crossing the Line between 'Straight' and 'Gay' in the US Navy* (New York: Routledge, 2011).

7 Paul Jackson, *One of the Boys: Homosexuality in the Military During World War II* (Montreal: McGill University Press, 2004); Emma Vickers, *Queen and Country: Same-Sex Desire in the British Armed Forces, 1939–45* (Manchester: Manchester University Press, 2013); Stephen Bourne, *Fighting Proud: The Untold Story of the Gay Men Who Served in Two World Wars* (London: I.B. Taurus, 2017).

8 Kellie Wilson-Buford, *Policing Sex and Marriage in the American Military: The Court-martial and the Construction of Gender and Sexual Deviance, 1950–2000* (Lincoln, NE: University of Nebraska Press, 2018).

9 Mike Hippler, *Matlovich, the Good Soldier* (Boston: Alyson Publications, 1989); E. Lawrence Gibson, *Get Off My Ship: Ensign Berg vs the US Navy* (New York: Avon Books, 1978); Stephen Snyder-Hill, *Soldier of Change: From the Closet to the Forefront of the Gay Rights Movement* (Lincoln, NE: University of Nebraska Press, 2014); James Lord, *My Queer War* (New York: Farrar, Straus & Giroux, 2010); Rich Merritt, *Secrets of a Gay Marine Porn Star* (New York: Kensington Books, 2005); Craig Jones (ed.), *Fighting with Pride: LGBTQ in the Armed Forces* (Barnsley, UK: Pen and Sword Military, 2019); James Wharton, *Out in the Army: My Life as a Gay Soldier* (London: Biteback Publishing, 2014).

10 Kristin Beck and Anne Speckhard, *Warrior Princess: A US Navy Seal's Journey to Coming Out Transgender* (McLean, VA: Advances Press, 2013).

11 Caroline Paige, *True Colours: My Life as the First Openly Transgender Officer in the British Armed Forces* (London: Biteback Publishing, 2017).

12 Hadrian, 'Homosexuality in the forces', *Campaign Circular* (November–December 1972), reprinted in *CAMP Ink*, issue 3 (1973): 4.

13 Ruth Ford, 'Lesbians and loose women: Female sexuality and the women's services during World War II', in *Gender and War: Australians at War in the Twentieth Century*, ed. Joy Damousi and Marilyn Lake (Cambridge: Cambridge University Press, 1995), pp. 81–104.

14 Ruth Ford, 'Disciplined, punished and resisting bodies: Lesbian women and the Australian armed services, 1950s/60s', *Lilith*, no. 9 (1996): 53–77.

15 Yorick Smaal, *Sex, Soldiers and the South Pacific, 1939–45: Queer Identities in Australia in the Second World War* (London: Palgrave Macmillan, 2015). See also Yorick Smaal and Graham Willett, 'Eliminate the "females": The New Guinea affair and medical approaches to homosexuality in the Australian Army in the Second World War', in *The Pacific War: Aftermaths, Remembrance and Culture*, ed. Christina Twomey and Ernest Koh (Abingdon, UK, and New York: Routledge, 2015), pp. 233–50; Graham Willett and Yorick Smaal, '"A homosexual institution": Same-sex desire in the army during World War II', *Australian Army Journal* 10, no. 3 (2013): 23–40.

16 Roderic Anderson, *Free Radical: A Memoir of a Gay Political Activist* (Salisbury, Qld: Lulu, 2006).

17 Noel Tovey, *Little Black Bastard: A Story of Survival* (Sydney: Hodder Headline, 2005), pp. 139–47.

18 David C. Collyer, *From Fishnets to Foxholes and Back Again* (Ballarat, Vic: David C. Collyer, 2001); David Bradford, *The Gunners' Doctor: Vietnam Letters* (North Sydney: Random House, 2007); *Tell Me I'm Okay: A Doctor's Story* (Clayton, Vic: Monash University Publishing, 2018).

19 Lorenzo Montesini, *My Life and Other Misdemeanours* (Ringwood, Vic: Viking, 1999), p. 165.

20 Katherine Cummings, *Katherine's Diary: The Story of a Transsexual*, rev. edn (Tascott, NSW: Beaujon Press, 2007), pp. 53–9.

21 Peter Stirling, *So Different: An Extraordinary Autobiography* (Sydney: Simon & Schuster, 1989).

22 Noah Riseman, 'Outmanoeuvring Defence: The Australian debates over gay and lesbian military service, 1992', *Australian Journal of Politics and History* 61, no. 4 (2015): 562–75; 'Transgender policy in the Australian Defence Force: Medicalization and its discontents', *International Journal of Transgenderism* 17, no. 3–4 (2016): 141–54; Shirleene Robinson, 'Sexuality at a cost: Lesbian servicewomen in the Australian military, 1960s–1980s', in *Beyond Combat: Australian Military Activity Away from the Battlefield*, ed. Tristan Moss and Tom Richardson (Sydney: NewSouth, 2018), pp. 45–58; 'A military discharge certificate' in *Queer Objects*, ed. Chris Brickell and Judith Collard (Dunedin: Otago University Publishing, 2019), pp. 297–301; Noah Riseman, '"Just another start to the denigration of Anzac Day": Evolving commemorations of LGBTI military service', *Australian Historical Studies* 48, no. 1 (2017): 35–51; 'Activism and Australia's ban on gay, lesbian and bisexual military service in the 1970s–80s', *Australian Feminist Studies* 33, no. 95 (2018): 147–63; 'From witch-hunts to pride balls: The ADF and LGBTI service personnel', in *Beyond Combat: Australian Military Activity Away from the Battlefield*, ed. Tristan Moss and Tom Richardson (Sydney: NewSouth, 2018), pp. 29–44.

23 Noah Riseman, Shirleene Robinson and Graham Willett, *Serving in Silence? Australian LGBT Servicemen and Women* (Sydney: NewSouth, 2018).

24 Carl von Clausewitz, *On War*, trans. Michael Howard and Peter Paret (Princeton, NJ: Princeton University Press, 1976), p. 76.

25 Morris Janowitz, 'Military institutions and citizenship in Western societies', *Armed Forces and Society* 2, no. 2 (1976): 185–204.

26 Warren L. Young, *Minorities and the Military: A Cross-national Study in World Perspective* (Westport, CT: Greenwood Press, 1982); Ronald R. Krebs, *Fighting for Rights: Military Service and the Politics of Citizenship* (Ithaca, NY: Cornell University Press, 2006).

27 John Maynard, 'The rise of the modern Aboriginal political movement, 1924–39', in *Serving Our Country: Indigenous Australians, War, Defence and Citizenship*, ed. Joan Beaumont and Allison Cadzow (Sydney: NewSouth, 2018), pp. 113–34; Riseman, 'Ex-service activism after 1945', pp. 261–81.

28 See Noah Riseman, 'LGBTI military service members and their partners', *Vida! Blog of the Australian Women's History Network*, 10 September 2017, http://www.auswhn.org.au/blog/lgbti-military-service/, retrieved 1 February 2019.

29 Martin Crotty and Mark Edele, 'Total war and entitlement: Towards a global history of veteran privilege', *Australian Journal of Politics and History* 59, no. 1 (2013): 15–32.

30 Marilyn Lake and Henry Reynolds, *What's Wrong with Anzac? The Militarisation of Australian History* (Sydney: UNSW Press, 2010).

31 Graham Seal, *Inventing Anzac: The Digger and National Mythology* (St Lucia, Qld: University of Queensland Press in association with the API Network and Curtin University of Technology, 2004), p. 10.

32 See Carolyn Holbrook, *Anzac: The Unauthorised Biography* (Sydney: NewSouth, 2014).

33 James Brown, *Anzac's Long Shadow: The Cost of Our National Obsession* (Collingwood, Vic: Redback, 2014).

34 Joshua Goldstein, *War and Gender: How Gender Shapes the War System and Vice Versa* (Cambridge: Cambridge University Press, 2001); Cynthia Enloe, *Maneuvers: The International Politics of Militarizing Women's Lives* (Berkeley, CA: University of

California Press, 2000); Teemu Tallberg and Johanna Valenius, 'Men, militaries and civilian societies in interaction', *Norma: Nordic Journal for Masculinity Studies* 3, no. 2 (2008): 85–98.

35 Wilson-Buford, *Policing Sex and Marriage in the American Military*, p. 39.

36 Jyonah Jericho, 'Hegemonic masculinity in the Australian Defence Force—the exclusion of women from combat service as state policy, 1973–2013' (PhD thesis, University of Sydney, 2015); Ben Wadham, 'Mogan hunts and pig nights: Military masculinities and the making of the arms-corps soldier' (paper presented at Australian Sociological Association Conference, La Trobe University, Melbourne, 2004) ('Mogan hunts' and 'pig nights' refer to evenings when male soldiers would get drunk, go out on the town and try to pick up the ugliest girl they could find.); Katerina Agostino, 'Men, identity and military culture', in *Male Trouble: Looking at Australian Masculinities*, ed. Stephen Tomsen and Mike Donaldson (North Melbourne, Vic: Pluto Press, 2003), pp. 108–31.

37 Ben Wadham, 'Brotherhood: Homosociality, totality and military subjectivity', *Australian Feminist Studies* 28, no. 76 (2013): 221.

38 Charles Moskos Jr, 'From citizen's army to social laboratory', *Wilson Quarterly* 17 (1993): 83–94.

39 Hugh Smith, 'The dynamics of social change and the Australian Defence Force', *Armed Forces and Society* 21, no. 4 (1995): 532.

40 Ben Wadham, Donna Bridges, Anuradha Mundkur and James Connor, '"War-fighting and left-wing feminist agendas": Gender and change in the Australian Defence Force', *Critical Military Studies* 4, no. 3 (2018): 271.

41 Rebecca Jennings, *Unnamed Desires: A Sydney Lesbian History* (Clayton, Vic: Monash University Publishing, 2015).

42 Trans-historicities look to the past, before the concept of 'transgender' existed, and use examples of gender non-conformity to imagine trans possibilities. Trans-historicities are the subject of a special issue of *Transgender Studies Quarterly* 5, no. 4 (November 2018).

43 Noah Riseman, 'The interview–document nexus: Recovering histories of LGBTI military service in Australia', *Archivaria* 87 (2019): 6–33.

44 Elizabeth Lapovsky Kennedy and Madeline D. Davis, *Boots of Leather, Slippers of Gold: The History of a Lesbian Community* (New York and London: Routledge, 1993).

45 Elizabeth Lapovsky Kennedy, 'Telling tales: Oral history and the construction of pre-Stonewall lesbian history', in *The Oral History Reader*, ed. Robert Perks and Alistair Thomson (London: Routledge, 1998), p. 348.

46 Horacio N. Roque Ramirez and Nan Alamilla Boyd, 'Introduction: Close encounters. The body and knowledge in queer oral history', in *Bodies of Evidence: The Practice of Queer Oral History*, ed. Nan Alamilla Boyd and Horacio N. Roque Ramirez (Oxford: Oxford University Press, 2012), p. 12.

47 See for instance Lynn Abrams, *Oral History Theory* (London and New York: Routledge, 2010).

48 Penny Summerfield, 'Culture and composure: Creating narratives of the gendered self in oral history interviews', *Cultural and Social History* 1, no. 1 (2004): 69.

49 Alistair Thomson, 'Anzac memories: Putting popular memory theory into practice in Australia', in *The Oral History Reader*, ed. Robert Perks and Alistair Thomson (London and New York: Routledge, 2006), p. 245.

50 Summerfield, 'Culture and composure', pp. 92–3.

51 Nan Alamilla Boyd, 'Who is the subject? Queer theory meets oral history', *Journal of the History of Sexuality* 17, no. 2 (2008): 188.

52 Thomson, 'Anzac memories', pp. 244–54; *Anzac Memories: Living with the Legend*, 2nd

edn (Clayton, Vic: Monash University Publishing, 2013); 'Anzac memories revisited: Trauma, memory and oral history', *Oral History Review* 42, no. 1 (2015): 1–29.

53 Popular Memory Group, 'Popular memory: Theory, politics, methods', in *The Oral History Reader*, ed. Robert Perks and Alistair Thomson (London and New York: Routledge, 2006), p. 46.

54 Wolf Kansteiner, 'Finding meaning in memory: A methodological critique of collective memory studies', *History and Theory* 41, no. 2 (2002): 187–8; Anna Green, 'Individual remembering and "collective memory": Theoretical presuppositions and contemporary debates', *Oral History* 32, no. 2 (2004): 37–8; Summerfield, 'Culture and composure', p. 34.

55 Abrams, *Oral History Theory*, p. 69.

56 Green, 'Individual remembering and "collective memory"', p. 42. See also Abrams, *Oral History Theory*, p. 30.

57 Kansteiner, 'Finding meaning in memory', pp. 187–8.

58 This is starting to change thanks to this project. See Melissa Davey, '"Injustice was official": Call for national apology to gay former military personnel', *Guardian*, 26 September 2018, https://www.theguardian.com/australia-news/2018/sep/26/injustice-was-official-call-for-national-apology-to-gay-former-military-personnel, retrieved 2 October 2018; Melissa Davey, '"I was absolutely shattered": The woman driven out of Australia's defence force', *Guardian*, 28 September 2018, https://www.theguardian.com/australia-news/2018/sep/27/i-was-absolutely shattered-the-woman-driven-out-of-australias-defence-force, retrieved 2 October 2018.

59 Boyd, 'Who is the subject?', pp. 182–3.

60 Trevor Lummis, 'Structure and validity in oral evidence', in *The Oral History Reader*, ed. Robert Perks and Alistair Thomson (London and New York: Routledge, 2006), p. 258.

Chapter 1

Hidden gay and bisexual men, 1944–73

In late 1943, authorities from the US Army informed their Australian counterparts that an investigation had identified almost fifty Australian soldiers who were allegedly involved in homosexual acts. The Australian commanders knew that they had to deal with this, but they had a problem: there was no clear policy about how to manage homosexuals.[1] This is not to say that the Australian Army had never grappled with the presence of homosexuals. Peter Stanley has identified cases of men charged with disciplinary action in World War I for homosexuality,[2] and there is at least one identified case of an Australian soldier charged with 'sodomy' against an eleven-year-old in Britain.[3] Tabloid newspapers like *Truth* occasionally featured stories about servicemen charged or convicted in civilian courts for acts of gross indecency.[4] Commanding officers could prosecute homosexuality under broad rules such as 'disgraceful conduct of a cruel, indecent or unnatural kind' or 'conduct to the prejudice of good order and military discipline'.[5] Both Yorick Smaal and Garry Wotherspoon also provide numerous examples of World War II servicemen participating in kamp subcultures (to use the parlance of the era) in home front cities.[6]

The Navy has a more transparent paper trail showing prosecutions for homosexual acts. The annually published *Return of Naval Courts-Martial and Disciplinary Courts* shows that nine sailors were charged for sodomy, acts of gross indecency or indecent assaults between 1916 and 1945, only two of whom were acquitted.[7] A 1939 naval document even expressed concern that some men were self-confessing to being 'homo-sexualists' to secure their own discharge. The memo recommended that such discharges should be classified as 'unsuitable' and that the dishonourable 'services no longer required [SNLR]' be attached to them. This would both punish those undesirable men and deter heterosexual men from feigning homosexuality to discharge from the Navy.[8]

The revelations from New Guinea in late 1943 seemed more serious. Yorick Smaal and Graham Willett have found examples of World War II commanding officers being relatively tolerant of homosexuality, so long as the men were discreet. Officers even tolerated gender non-conforming practices so long as they happened within particular bounds, such as drag shows for troop entertainment.[9] Now that the Americans raised the issue of homosexuality, discretion was gone, and the Australian Army needed to deal with the problem of homosexuals in its ranks. For nine months, Army officials deliberated the formulation of a policy on homosexuality, presenting the first formal directive on homosexuality in 1944. In cases involving minors, violence or public obscenity, commanding officers should take disciplinary action against homosexuals. Other cases should be treated as medical issues, with psychological evaluations guiding whether the member could be 'treated' and retained in the Army. If they could not, then they would be discharged on medical grounds.[10]

The World War II precedents, both in policy and in practice, laid the foundations for how the Australian services dealt with male homo/bisexuality in the post-war era. As historians such as Graham Willett, Garry Wotherspoon, Robert Reynolds and Shirleene Robinson have written, the 1950s and 1960s were a time when there was much public silence surrounding homosexuality. That could be good for men who were inconspicuous, but it posed challenges for those wanting to be openly gay/bisexual or for those who were not discreet. Gay and bisexual men sometimes cruised for sex in public places known as beats; in fact, Melbourne's Shrine of Remembrance was one known beat during and after the war.[11] Those who were caught faced stiff penalties, and police were known to entrap gay/bisexual men. The topic of homosexuality appeared in the media mostly through published accounts of police actions or court reports, and there was a marked increase in convictions for homosexual acts in the 1950s.[12]

In the armed forces, too, the 1950s and 1960s were a time of silence around male homosexuality. Whereas women who joined the military in the 1950s and 1960s transcended expected constructs of femininity and therefore came under greater scrutiny, men who joined the services were fulfilling cultural expectations of masculinity. Silences were partly reflective of the fact that male involvement in the military was considered both desirable and patriotic. The silences around homosexuality meant that service police gave little thought to the presence of gay or bisexual men, so the practice of hunting homosexual men was not common. Oral histories from gay and bisexual ex-servicemen reinforce the notion that silences worked in their favour: many recall same-sex activity happening regularly on bases or ships, and it being a non-issue. Fellow servicemen regularly knew or suspected certain men to be gay or bisexual, and they were happy to overlook it or sometimes even to approach them for sexual favours. It was only when someone became too open—whether caught in public or attracting attention from someone more hostile to homosexuality—that authorities intervened. Even then, the absence of a strong policy, and military desires to keep the issue of homosexuality quiet, meant that authorities dealt with these gay or bisexual men discreetly, letting the cycle of silence continue unchallenged.

This chapter explores the many complexities around silences and the experiences of gay and bisexual men from the end of World War II until the early 1970s. Importantly, the focus of this chapter is on men. As chapter 2 explores, lesbian and bisexual servicewomen faced a much more hostile environment during this era. This chapter first explores the challenge of gay/bisexual men's own identity constructions. Silences around homosexuality and the enduring power of heterosexism—the assumption of heterosexuality as 'normal' and other sexualities being deviations—often made it difficult for men to realise, articulate or explore their sexual orientation. The Vietnam War experiences of gay/bisexual men are an interesting microcosm of the entire era because there was little homosexual activity, often because servicemen did not yet realise they were gay or because there were few same-sex opportunities. Yet for those who did have the opportunity, authorities often turned a blind eye to homosexual acts. This practice of tolerating homosexuality, so long as the men were discreet, continued on Australian ships and soil. The chapter finally turns to what happened when men were not discreet, showing the ways authorities policed homosexuality. The lack of coherent policies meant that investigations and discharges were inconsistent both within and across the services, just as homosexuality was emerging as a topic of public interest.

The archetypes

Silences among gay and bisexual men from the 1950s and 1960s could endure for decades, with many men denying, repressing or hiding their sexuality and never living an openly gay life. The passing of time makes it harder to uncover stories of homosexuality in the immediate post-war period, as many gay or bisexual servicemen from that era have since died. Nonetheless, some openly gay men who served in that era, or the families of deceased veterans, have shared stories. The gay and bisexual men from that era loosely fall into three archetypes: those who embraced their sexuality and pursued same-sex encounters; those who did not realise their sexuality until later in life, after they discharged; and those who knew they were gay or bisexual but tried to sublimate it, either by having relationships with women or by practising celibacy (mostly).

Some of the gay men who accepted their sexuality had an aura of boldness, engaging in what one might consider risky behaviour given the public taboos around homosexuality. Graham Jamieson, widower of Tom Goldsby, relates stories that Goldsby used to tell about his time in the British Army (1947–49; 1950–52) then the Australian Army (1952–58). One anecdote from the British Army in the Korean War involves Goldsby calming down a nervous comrade in the trenches by having sex with him. In a more feasible but equally bold tale, Jamieson describes when Goldsby was on sentry duty in Australia and a messenger arrived at the base. Goldsby said to his mate: '"You cover for me. I'm going to go back and fuck this bloke." Tom said when he got to the bloke's room, the bloke says, "Make sure you take your boots off. They're clean sheets."' From 1955 to 1957, Goldsby was stationed in Penang and conducted regular patrols and ambushes during the Malayan Emergency. Goldsby forged a relationship with a local Malay man, who embarrassed Goldsby by dedicating a song to him over the radio. Jamieson does not know how Goldsby explained that away to his mates and just speculates that he 'deafed it'.[13]

One reason Tom Goldsby got away with his bold behaviour is that, by all accounts, his peers liked him. In 1972, 'Hadrian' wrote about his fourteen years in the Army, including service in New Guinea, Borneo and Korea. During his career he had numerous conversations about homosexuality with other soldiers. He believed that 'the average soldier couldn't care less' and that soldiers were more prone to judge each other by their ability. Hadrian recalled only one man who was discharged for homosexuality, but attributes that to the man being 'universally disliked'. When he was caught and discharged for homosexuality, 'Everybody cheered like mad, not because the WO [warrant officer] was camp, but because he was such an utter bastard.'[14]

Tom Goldsby was not the only Australian to have a same-sex relationship in Malaya. Bob Flanagan had joined the Army in late 1959 and served in Melbourne for two years, where he was already having sexual relations with men. In 1961 Flanagan began a two-year deployment to Taiping and then Seremban as one of sixteen Australians serving alongside about a hundred British soldiers. One weekend night in Kuala Lumpur, Flanagan went to a bar-café which hosted dances that catered to a gay and gay-friendly clientele. He met a working-class Malay man on a motorbike and began a relationship that lasted for a few months. The gentleman even introduced Flanagan to his family as his Australian 'mate'. Flanagan then met another gay Malay, this one a wealthy owner of an orchid farm. That relationship lasted only a few weeks because an Australian soldier-mate found out and became jealous. As Flanagan recalls, the mate 'came up and threatened to knock the shit out of the bloke I was dancing with, and that's when he took me straight back to Seremban. We booked into a hotel and we had sex.' The other man was a cook and had access to his own room, so for the next eighteen months he and Flanagan had a relationship.[15]

Bob Flanagan does not know if the other Australian soldiers ever knew about the relationship, although he suspects they did not because he believes they would have 'probably belted the shit out of us or something like that'. The British, however, did catch Flanagan and the cook in bed one time; he recalls: 'They didn't act shocked; they were quite comfortable with us being in bed together—"Come on, you bastards, get out of bed and let's play cards." You know, just like, "Finish your breakfast", or something like that. So I found their attitude was different to the Australians.'[16] Flanagan's worries about the Australians suggests, from his perspective, a more open culture of tolerance in the British Army. That said, given the close quarters, the nature of the relationship and the fact that the British soldiers caught Flanagan and the cook, it is quite likely that some of the other fourteen Australians either suspected or knew about their relationship. A key factor here that other cases in this chapter reinforce is that when servicemen were discreet, it was easy to turn a blind eye.

Brian Allen served in the British Army from 1953 to 1956 and subsequently emigrated to Australia, serving in the Royal Australian Electrical and Mechanical Engineers from 1959 to 1980. In both armies he used to 'fool around' with civilians at beats, and the occasional soldier as well. Allen did a stint with a UN peacekeeping force in South Korea in 1965, where he even picked up a Korean man. At numerous stages throughout his career, especially in the 1970s, Allen had civilian boyfriends, some living with him. By the 1970s Allen was even frequenting gay bars in Melbourne. When asked how he kept this a secret from the Army, Allen was clear that 'in those days it wasn't an issue. It's a question of, if you're not asked, you don't tell.'[17] Although Allen was anachronistically referring to the US policy of 1993–2011, the key point he was making was that male homosexuality was not on the military's agenda. For men such as himself, Bob Flanagan and Tom Goldsby who were discreet, there was scope to have same-sex relations and not draw the attention of military authorities. Indeed, in 1972 Hadrian recalled a conversation with a straight major about the discharge of a British officer caught in bed with a national serviceman: 'The Major said simply: the stupid bugger—he could have been a bit more discreet about it—with cheap hotels all over the place, why did they have to pick on the barracks to have a nooky.'[18]

Trevor Robinson is a good example of the second category of men, who did not realise their sexuality. Robinson served from 1952 to 1967, including almost going off to the Korean War (but an injury left him in Japan) and participating in the 1956 Maralinga nuclear testing. Robinson recalls that during his Army years, he had no consciousness or understanding of homosexuality, even as he came across other gay or bisexual soldiers. He recalls one incident when a sergeant came to his room in Puckapunyal:

> He obviously was trying to get a bit fresh with me and I didn't know what he was up to. I had no idea that you'd put an erect penis up somebody's ass. No idea. Simply no idea. So he's playing around with me and I hit, I pushed him aside and he hit his head on the fireplace in my room.

Robinson recalls that the sergeant and several other gay or bisexual officers and NCOs had wives to disguise their homosexuality (or 'beards', to use a present-day colloquialism).[19] By his own admission Robinson was naïve and ignorant about homosexuality, but it was also fear about what might happen to outed homosexuals that prevented him from realising his true sexual desires.

Fear often motivated the third category of men who recognised their homosexuality but remained celibate. The fear was sometimes about what would happen if they were caught by military or civilian authorities, but several interviews reveal another influence: religion, and especially Catholicism. Tony Whelan was raised a strict Catholic and had even been involved in the Catholic-dominated Democratic Labor Party before being called up for National Service from 1969 to 1971. Whelan resisted acting on his homosexuality until later years, although he jokes about his time in the Army: 'There were a couple around there who lived in the barracks who I would have liked to have known, particularly a beautiful blonde who used to pull apart minis for some reason! I'd have liked to have pulled some of his clothes apart, but it never happened.'[20] David Morrison (who shares the same name as the former Chief of Army) was a flight sergeant who served in the RAAF from 1952 to 1972 as an aircraft instrument fitter. Morrison had a strong Catholic upbringing, and this drove him deep into the closet. Morrison recalls being propositioned at least once by another airman, but when asked if he reciprocated, he responded: 'No, I couldn't; the Church and the Air Force. I was clamped.' Morrison even married while he was serving. He never came out to his wife and by the 1980s began secretly meeting other men for sex, but he continued to be a devoted husband until his wife died in 2007.[21]

Brian McFarlane, too, was raised a strong Catholic, and this—combined with a low libido—strongly influenced his behaviour as a closeted bisexual man during his service from 1951 to 1975. McFarlane's long career included stints as a trainer at the Royal Military College, Duntroon and in Malaya and

Vietnam. During his career he witnessed many other gay and bisexual officers whose sexuality was known but tolerated. He specifically notes:

> One lieutenant colonel, the commanding officer of the field artillery regiment in Vietnam, was gay, although of the old style who would not have wanted all and sundry to know, but they did. A lieutenant colonel commanding the artillery field regiment in Singapore at one stage was gay. He was a most personable and very well-liked man, who I think did not care who knew, and everybody did. Post his time in the Army, he set up a restaurant with his boyfriend. Both those officers were single. There was another officer I have not previously mentioned from Malaya days, married, but gay, who later rose to be Director General of his corps. There was also a single man who was gay and was the Director General of his corps.[22]

From the commencement of his time in the Army until the late 1960s, McFarlane had sexual relations with a man only once, and this was a civilian. A pivotal moment transformed McFarlane into a much more confident leader who was more willing to take risks later in his career: the Vietnam War.

Vietnam War

In 1970, Dr Gavin Hart, a registrar at the First Australian Field Hospital in Vung Tau, conducted a study investigating the sexual behaviour of 670 Australian servicemen in Vietnam. Only fourteen of the survey respondents—or 2 per cent—reported having homosexual experiences in Vietnam. Hart said that this was a statistically insignificant number, although he did note that four of the men claimed to be happily married; eight reported having intercourse, one fellatio and five masturbation. Hart also concluded: 'In a confined barracks area, with its lack of privacy, the risk of disclosure of any homosexual approach or union was too great for most homosexuals to accept.'[23] Oral histories with Australian Vietnam veterans reinforce this finding, but they also reveal that the attitudes, coping mechanisms, behaviours and experiences of gay and bisexual servicemen in Vietnam were more complex.

The three archetypes of gay/bisexual servicemen continued to serve in the forces during the Vietnam War. All three types faced challenges because of the close environment, where they were in constant contact with other servicemen. Even celibate men faced difficulties because so much bonding, particularly during R&C (rest and convalescence) leave in Vung Tau, was framed around sex. Vung Tau was a site of rampant drinking, brawls with local

Vietnamese and American servicemen, and relations with Vietnamese sex workers (which contributed to high rates of venereal disease among Australian servicemen).[24] Brian Allen evaded Vung Tau by spending his R&C time with three mates teaching English at a local school.[25] Dr David Bradford, at the time a religious member of the Uniting Church, avoided visiting Vung Tau.[26] Other gay men did go to Vung Tau and did partake in the sex trade with women. Alan Lewis states: 'I went to Vung Tau, went through all the bars. I had sex several times with bar girls but yeah, I was kind of—I don't know that I was gay. I mean I was gay obviously, you know, but I wasn't thinking about it, you know. I wasn't sort of thinking much about it.'[27] 'Peter', who was a national serviceman in 6th Battalion, Royal Australian Regiment (6RAR) from May 1969 to May 1970, recalls struggling with his sexuality at the time. As he explains: 'Vietnam was probably the hardest period when I was trying to work out, you know. Yeah, I suppose, I was having this incredibly intense attraction to males, and yet when we went into Vung Tau, Yvonne, the beautiful French Vietnamese prostitute—I had an incredibly strong attraction to her. So, yeah, and so things were difficult in the Army.'[28]

The circumstances in Vietnam meant that options for homosexual acts were limited, especially for the majority of soldiers who served at the 1st Australian Task Force base at Nui Dat. Being on active deployment brought men into even closer proximity to each other. There were heightened levels of stress but also a strong sense of camaraderie. Friendships became emotionally intimate because of the shared experience of danger, isolation and loneliness. Sebastian Junger describes the tight emotional bonds among servicemen forged in wartime as a tribal connection.[29] One might expect that such physical closeness—not to mention the homoerotic environment with so many fit, bare-chested men walking around, often naked—to be a hotbed of same-sex encounters.[30] Yet this was not the case. If anything, the heightened stress and close quarters served more as a form of surveillance, with few men having the privacy or opportunity to act on any same-sex attraction. Indeed, 'Peter' used to shower early in the morning to avoid seeing other soldiers for fear of getting an erection.[31]

For many gay or bisexual men, celibacy was not a choice but a situational imposition. In a 1988 article in gay magazine *OutRage*, Vietnam veteran Bill commented: 'In Vietnam you had no opportunity to be alone with anyone. Our time was controlled. No privacy whatever.'[32] Gay interview participants such as Brian Allen, Alan Lewis, 'Peter' and author David Collyer had regularly visited beats in Australia. In Vietnam, they did not have any sexual relations with other men. Partly this was because of the nature of the job and the 24/7 combat environment, but it also had to do with lack of opportunity. When asked about sex with men in Vietnam, Alan Lewis definitively answered: 'No.

No, never ever, ever. No, there were no gays, no poofters back then in the Army, no. Not that we knew of.'[33] David Collyer writes:

> I can say that during the times we were on operation, the thought of having a wank never even entered my mind. I have heard it said that you could not have gay men on the front lines because they would be forever attempting to seduce the straight blokes, or at least to put the hard word on them. This is absolutely ridiculous! I knew I was gay, and I knew they were straight. The two just don't mix sexually.[34]

There were, of course, some situations when gay men did have opportunities with each other. These tended to arise when someone was in a position of authority and had access to private space. Dr David Bradford was regimental medical officer to the artillery's 4th Field Regiment from May 1967 to May 1968. He remembers one Sunday evening when the duty non-commissioned officer (NCO) came to him for advice. The regimental sergeant major had returned from the US NCOs' mess with another man. Bradford recalls that the duty NCO said:

> 'David,' he said, 'what am I going to do?' And I said, 'What, what about?' He said, 'The RSM has got a guy in his tent and they're having sex; I can hear them.' And I said, 'Well, why is it your business?' And he said, 'I'm on duty, so you know this is a military offence.' I said, 'Stand down afternoon [*sic*]; don't be so ridiculous.' I said, 'How are you going to get on if you do something and report him—he's a regimental sergeant major.' I said, 'You know your superior. How are you going to get on if you know there's absolutely no proof that he is doing anything sexual with this boy?' He said, 'So you think I shouldn't do it?' I said, 'You're mad if you do do anything.'

Bradford observed that everyone knew about the regimental sergeant major, but the reaction was 'most people sort of just shrugged their shoulders a bit, you know. Whatever makes him happy, you know.'[35]

Brian McFarlane says he knows of nine gay or bisexual men, all officers, who served in Vietnam. There were certain officers whom everyone at Task Force headquarters knew were gay, yet the subject never came up. He explains: 'There were always gays, both male and female, in society and in the military. Those closely associated with them mostly always knew of their inclination. But the sensible gay did not go about pushing the matter in the faces of all and sundry, particularly those who may not understand, or care to understand.'[36] Bradford and McFarlane's examples signal that rank provided a shield to protect gay men from being targeted or kicked out. Partly this was because lower ranks would not challenge them, but there also seems to have been more respect for their work and effectiveness as commanders and leaders. This, of course, assumed discretion among the gay officers and NCOs.

Whereas the oral histories suggest that Nui Dat was a generally sexless place, gay and bisexual men who served at the Australian Logistic Support Base in Vung Tau tell a different story. This was the site of the main hospital, engineers, ordnance, transport and the RAAF units. National serviceman Lorenzo Montesini served there as a medic from April to December 1967. According to Montesini, three nights before departing Vietnam, he and Dental Corps member Robert Straub got drunk and ran to the beach, where they connected in the water in a vision Montesini describes as something out of a romance movie. He continues:

> We stayed on the beach most of the night and this was my first physical, full physical experience and then we went back to the lines and he put me, it was just like a long hut with cots, and he put me to bed and then he went off and I just put my mosquito net down and I just rolled and went to sleep. And then when I woke up he was standing there and he said, 'I just wanted to make sure that you were here. I was here when you woke up because I wanted to tell you that last night was not just, "Oh sorry, mate", it was not that.'[37]

That night was the beginning of a tumultuous twenty-seven-year relationship that ended when Rob Straub died of AIDS-related causes in 1995.

Servicemen stationed at Vung Tau had more freedom of movement and access to the township and the many vices it offered. Randy Shilts's book *Conduct Unbecoming* briefly references Vung Tau as a common site for Australian homosexual encounters.[38] Only one Australian oral history suggests that there might be some truth to Shilts's assertion, albeit not in a positive way. RAAF Sergeant Leon Fry was an administrative clerk at the Australian Logistics Support Base from July 1970 to July 1971. Although Fry had a few sexual encounters with other airmen, he worried about getting caught and therefore preferred to look elsewhere. Fry used to slip over the fence and visit the US clubs in Vung Tau because 'there was always a bit of action there'. On 24 October 1970 (Fry clearly remembers the date), he paid a visit to a US club and met a hot gay soldier. He describes what happened next:

> He invited me around one night and I went around late and he said 'Well, we're closing.' I said, 'Oh, we're going to have a fuck, yeah, good fun, that's what I'm looking for', and there were a couple of other guys there. But what I didn't count on was getting raped by the three of them. That was fucking awful [pause]. I didn't know what it did. It shattered me. I thought I was going to go mad. Happened on a Saturday night, and the Sunday I wrote myself off totally. And Monday I went to work and my Admin O [Administrative Officer] said, 'Whatever's wrong with you?' I said, 'Oh, I got some bad news from home and wrote myself off.' He said, 'Well, you better have a day off.'[39]

The rape traumatised Fry and began a long cycle of depression and alcoholism that plagued him until he retired from the RAAF in 1989. His story reveals that sexual violence in the military also could affect men as victims rather than just perpetrators.

Other reported examples of homosexual relations in Vietnam tended to be in exceptional circumstances when service members had access to private space. National serviceman Bill in 1988 wrote about having sexual relations with another soldier when they were assigned a special mission that gave them a private hut:

> He was a man with an enormous sexual appetite, but in Vietnam he wouldn't go to places where you could meet bar girls. He was terrified of getting VD. But he wanted sex. He decided that in fact he had got sexual pleasure from his jail experience [before the war]. So sex with me became his only outlet. That's how it happened. For some months we had sex in the hut.[40]

Shilts writes about examples of heterosexual American soldiers sometimes propositioning men perceived to be gay; they reasoned that getting a blow job from a man did not make them gay, instead representing what psychologists call situational homosexuality: where men who identify as heterosexual will engage in same-sex activity when they are in a long-term environment (e.g. prison; military combat zone) where there are no women.[41] Interestingly, Shilts argues that straight guys were more willing to engage in such acts while gay servicemen were more likely to repress their sexuality for fear of being caught.[42] It is highly likely that there were other Australian cases of situational homosexuality in Vietnam, and several of the gay ex-service members mention it. After the war Brian McFarlane learned about a lieutenant colonel involved with a captain from another unit. McFarlane speculates that it might have been only a 'shipboard romance', and he explains: 'One of the reasons why these people would be assessed as being gay, or be getting off, is loneliness. They'd probably just want to be in the same bed with somebody, for the warmth of friendship. Not even necessarily be having it off. Probably never were. But I think, loneliness was the thing that used to get to people.'[43] Yorick Smaal's work on homosexual activity during World War II similarly found that the emotional intimacy among soldiers sometimes could 'rupture in particular moments of vulnerability or hardship'.[44] Hadrian testified before Western Australia's 1974 Honorary Royal Commission to Inquire into Matters Relating to Homosexuality:

> During my years in the Army in wartime [World War II and Korea] I performed sexual acts with many comrades. Some were confirmed homosexuals but most were engaged or happily married, but away from their wives and sweethearts. The first approach often came from them. Those men were able to gain relief from their sexual tensions among themselves and were none the worse for their experiences. I met many of them later and all considered these incidents of no consequence.[45]

Whether same-sex encounters involved men who identified as gay/bisexual, or were examples of situational homosexuality, discretion was still of the utmost importance or else men would face formal or informal sanction. Bill recalled in 1988: 'One particular night a group of them got high on grass and one of them said he gave better head than any bar girl. He apparently proved it and was caught. For the rest of his time in Vietnam he was harrassed [*sic*] dreadfully. I think the CO was happy to see the harassment as a deterrent to anyone else.'[46] In one case where authorities did take action, an officer awoke from a drunken party in February 1968 in the lock-up. After a thirty-two-hour interrogation, the chaplain told him that he was being charged for having committed homosexual acts. It was alleged that the man put his arm around another serviceman's waist and attempted to digitally penetrate him. Although the accused serviceman does not remember this and believes it to be false, there was a clear reason that authorities took action against him: he was neither discreet nor engaging in a consensual homosexual act.[47] Thus, even as the boundaries blurred between straight, gay and bisexual, and commanders could prove tolerant of homosexual activity, still public expressions of same-sex affection remained outside the bounds of acceptable behaviour.

Navy ships and sexuality

The Navy has a distinct Vietnam War history, primarily serving as transport ships for troops, but also participating in some gunfire operations in the Tonkin Gulf and mine-clearing operations.[48] Sailors who served on ships that travelled to Vietnam are also classified as Vietnam veterans, and their stories of homosexuality both during the war and during the period from the 1950s to the 1970s shed more perspectives on the complex dichotomies between gay or bisexual identity and homosexual behaviour. The Navy was also the only service to have clear policies about homosexuality in place, although the treatment of suspected gay and bisexual men in practice varied substantially.

Wally Cowin's experience epitomises the homosexual subculture lurking in the open on board ships. Cowin joined the Navy as a sixteen-year-old in 1965. By 1966 he was a kitchen hand on board HMAS *Sydney*, where he participated in four journeys to Vietnam between 1966 and 1968. *Sydney* ferried many nervous young soldiers to Vietnam, and Cowin certainly was willing to sway them towards sexual adventures: 'I had used that line on some of them too because sometimes it worked. "You have never had your cock sucked by a man and you're going to war? We'll fix that for you right here and now."' Cowin describes other same-sex traditions on ships, such as searching for the Golden Rivet. The mythology is that the final rivet nailed into a ship is made of gold, and new, uninitiated sailors are often tricked into searching for it. Cowin recalls that older gay or bisexual men would invite new sailors whom they suspected of homosexual inclinations to search for the golden rivet and, when they were in a secluded part of the ship, they would have sex.[49]

It is not only Cowin who recalls men who identified as straight who were more than happy to have sex with other men on board ships. In 1966 one sailor wrote in a police statement that he had masturbated one sailor on HMAS *Vendetta* and allowed another on HMAS *Melbourne* to suck his penis in exchange for beer.[50] 'Xavier' was a leading seaman who served from 1962 to 1974, including two trips to Vietnam in 1965 and 1969, and as a crew member on HMAS *Stuart* searching for survivors of the *Voyager–Melbourne* collision in 1964.[51] Xavier was deep in the closet because of his strong Catholic upbringing; he had a few homosexual encounters with civilians, but was terrified of sexual relations with other sailors. Even so, other sailors propositioned Xavier on more than one occasion. He recalls: 'One particular fellow always managed to be naked when I was. He would know when I was, on the *Melbourne* … And he was presenting himself to me, which I just thought, nah, I'm not going there. I knew he was married, and I thought, nup. Not doing that.'[52]

Xavier also notes that there was a dichotomy between those men who were known to be gay but tolerated, and those who identified as straight but enjoyed sexual relations with men. Xavier explains:

> Well, there was a couple of blokes who were stewards, and people used to think, 'Oh well, he's a steward, he looks after—he makes the beds and he looks after the officers and he's—and he services the dishes. He does feminine things. So, if he's queer, well, that's OK, so long as he does his job and doesn't touch anybody, or say anything to anybody that's fine.' There may have been others he was having relationships with but you'd never know about it.[53]

Allan Bérubé has similarly commented on the queering of occupations within World War II US forces. Because roles like stewards, chaplains, nurses or secretaries were not associated with dominant constructs of masculinity, they became a safe place for homosexual men to serve in an open closet.[54] John Dunlop was a divisional officer in the Navy from 1954 to 1974, and he recalls: 'Often, you'd find that some of the stewards or SBAs [sick berth attendants] would just be a little bit camp and everyone accepted [it]; it was just part of the scene. It didn't raise—it might have caused a bit of a chuckle sometimes, but that's all.'[55] As even Xavier learned years later, both straight and closeted men knew that they could approach these men for sex. Indeed, Xavier recalls running into an ex-sailor mate in a gay bar around the year 2000, and the man revealed that he had sexual relations with the steward on their ship.[56]

The tolerance shown to discreet gay sailors is all the more remarkable because the Navy was the only service between the 1950s and the 1970s that had clear policies for investigating homosexuality. From at least 1954 the RAN adopted policies from the British Royal Navy on 'unnatural acts': the charges of buggery, assault with intent to commit buggery, indecent assault, act of gross indecency with a male person, and procuring the commission of an act of gross indecency with another male person. The policy directed that any cases with clear evidence should go to a court martial. In cases where the evidence was less compelling or relied only on the word of other homosexuals, there was scope for commanding officers to give suspected homosexuals a warning. Commanding officers would still report these men to the Naval Board, which would keep a list to be consulted in the event of further charges.[57] 'Peter' remembers this practice still happening when he worked in the Navy legal office in 1976.[58]

The policy on unnatural offences outlined the procedures for investigating suspected cases of homosexuality. It placed significant weight on uncovering medical evidence to prove that same-sex activity had occurred. The naval instruction even authorised doctors to conduct invasive anal exams searching for 'Recent tears, lacerations, fissures and piles, old scars due to previous ulceration or any other sign that might be present and might cause dilation or relaxation of the anal sphincter'. Similarly, doctors should conduct penile examinations to search 'for evidence of friction, for the tearing of phrenum and presence of faeces especially beneath the prepuce if uncircumcised ... Examination of the base of the penis should be made for contamination with faeces and spermatozoa.'[59] None of the ex-sailors interviewed for this project had to undergo these medical tests, but there is evidence that such examinations occurred. A doctor provided evidence for a 1957 joint court martial of three sailors charged for indecent behaviour. He examined their external genitals, anal area and 'proctoscopic examination of anal canal and lower rectum', finding no evidence of 'trauma discharge' in any men (these men were found not guilty of the charges).[60]

In 1968 the Navy updated its rules to be titled 'abnormal sexual behaviour' instead of 'unnatural offences'. The policy retained the instructions about medical examinations, but the instruction gave clearer scope for commanding officers to take action other than courts martial. Where there was no clear medical evidence and the suspected homosexual came to the commanding officer's attention through a complaint or self-confession, the commanding officer could request an administrative discharge, but still with the dishonourable classification of 'SNLR [services no longer required]' or 'unsuitable'. The policy also, for the first time, made a distinction between so-called 'confirmed homosexuals' and those who had committed an 'isolated instance of homosexuality'. Reflecting ideas about situational homosexuality, the instruction indicated that commanding officers could, under the advice of medical officers and psychiatrists, elect to retain men who had committed homosexual acts if 'psychiatric treatment or merely a sympathetic understanding of the individuals problems will correct a potentially unhealthy situation'.[61]

It was primarily in cases when there had been a non-consensual act or when sailors were indiscreet that commanding officers acted on these policies. A good example of the latter again comes from Wally Cowin. His oral history includes numerous tales of sexual encounters with soldiers and sailors, but it was a final encounter at HMAS Watson in 1969 that proved to be his undoing. Cowin was working the kitchen during a dinner for the officers and NCOs. He went to the victualling office between courses, where another sailor was on the phone. The other sailor then undid Cowin's fly, and the two had sex. They were so involved in the act that they neglected to remember that the lights were on and several off-duty sailors could see the entire act through a window. As Cowin recalls, 'Then they sort of watched and watched and watched and watched, right to the very, very end. As we were sort of tidying up and getting ... then the phone rang again. It was somebody we both knew. "What? What?" ... Then we both sort of turned around and everyone ... "Ha, ha."'[62]

Even this blatant breach of discretion did not immediately result in disciplinary action. Cowin recalls that most other sailors joked about it after, but there was one sailor who disliked him and reported the incident. The commanding officer followed the instruction on abnormal sexual behaviour and elected to avoid the court martial, instead sending Cowin to a psychiatrist. The psychiatrist confirmed that Cowin was gay (and he says that the psychiatrist, too, was gay), and on 3 October 1969 Cowin received a dishonourable discharge 'services no longer required'.[63] Wally Cowin's story shows the numerous ways that authorities could exercise significant discretion when dealing with cases of homosexuality: they could turn a blind eye, although this was not possible when the breach was so public. Then faced with the decision to court martial or an administrative discharge, the commanding officer chose the latter. Such an approach mitigated the public embarrassment accorded to Wally and, depending on how an investigation might have played out, potentially to other sailors.

In another example, an accusation of sexual assault sparked an investigation into unnatural acts on HMAS *Melbourne* that led to the discharge of four sailors and two other men being placed on the Naval Board's list of suspected homosexuals. It began on 8 February 1966 when a sailor self-confessed to being gay but, even more than that, he named other sailors with whom he had sexual relations and others who were rumoured to be gay. The most chilling accusation, though, was that in February 1965 two other sailors had pinned him down and sexually assaulted him on board HMAS *Melbourne*. The investigation report suggested that the victim was 'apprehensive and close to an emotional breakdown', and he was sent to a naval hospital for psychiatric observation; his fate after that is unknown. The accused perpetrators never faced a court martial because there was no conclusive evidence of sexual assault. Nonetheless, the investigators determined that an incident of homosexuality had happened and recommended that the two alleged perpetrators, the victim and a fourth sailor who had been present be discharged 'services no longer required'. A letter to the Minister for the Navy justified this decision because 'the retention of any of these sailors could pose a security risk, in addition to the obvious desirability of removing those with homosexual tendencies from contact with other sailors in the RAN'.[64]

Investigators later questioned some of the other men whom the sailor accused of being gay, and two of them admitted to having taken part in homosexual acts in the past. Both were retained in the Navy, although one had his security clearance reduced, and both names were reported to the Naval Board. Another man named in the HMAS *Melbourne* investigation denied being a homosexual, but investigators saw his name on the Naval Board's list of previously accused. The Department of Naval Intelligence deemed him not credible and discharged him 'services no longer required'.[65] Hence, while the punishment for accused first-offender homosexuals would vary depending on circumstances, repeat offenders from the other ranks generally faced discharge.

Subcultures and protection

The investigation into HMAS *Melbourne* reveals other aspects of homosexual subcultures in the Navy. The witness and psychiatrist statements indicate that many sailors clearly knew there were homosexual subcultures on board ships. The consultant psychiatrist John McGeorge, on examining one of the four men set for discharge, wrote, 'His claim that he is shunned seems rather exaggerated for many homosexuals in the service do not seem to have been treated similarly … I have grave doubts as to the truth of his story that he was "hounded". I have never known it to happen to other homosexuals and I cannot see why he should be the exception.'[66] The key point to take from Dr McGeorge is that the norm was *not* abuse of gay/bisexual men but rather tolerance. One of the other men caught in the investigation reported that 'prostitution was endemic among sailors as a source of easy money'. He named the Quarterdeck Bar at the Chevron Hotel in Kings Cross, Sydney, as a place frequented by homosexuals and where sailors would often make extra money by meeting men who would pay for sex.[67] Such statements suggest the existence of subcultures among gay and bisexual sailors, with their own drinking haunts and social cues.

Just the presence of one known gay or bisexual man could be enough to stir the sexual interests of other men. Richard Williams was a corporal medic serving with 1st Battalion, Royal Australian Regiment in Townsville when his sexuality first came under suspicion in late 1972 or early 1973. The commanding officer summoned Williams to advise him that people had seen him sitting on their beds and making lewd comments and thought he might be gay. Williams denied the allegations, and the commanding officer cautioned: 'Just be very careful in what you're doing because if this happens again, you'll be discharged.' Word spread about Williams being a suspected homosexual; rather than leading to any ostracism or bullying, he was inundated with sexual offers. For instance, he recalls 'one guy [who was] a very macho rough guy who was another corporal in Delta Company, that used to sneak up to my room when he'd had a load of booze and wanted to have sex with me and I didn't particularly want to have sex with him. But he was a lot stronger with me so that happened.' Williams began having sexual relations with other soldiers as well. Unfortunately, one of these other men was, as Williams puts it, 'A little bit more prolific in his putting the hard word on people. And he put the hard word on one too many who made a complaint to the military police.' During his police interview, that corporal dobbed in Williams, who was subsequently discharged from the Army for homosexuality in 1974.[68]

Garry Gleadhill recalls a strong homosexual subculture among other ranks while he was working at the officer training base HMAS Creswell from 1968 to 1970. Gleadhill recalls one incident when a steward caught him with another man in his room. He describes the response: '"For Christ's sake do it ashore, mate, don't do it on board because if somebody catches you and tells an officer, you're gone, mate. Goodbye career." And that was it. Lift the carpet, quick sweep, all gone under the carpet.' As Gleadhill told more stories about HMAS Creswell, he was describing not just a homosexual subculture but also a culture of protection. He says, 'You knew exactly what was happening. People would go off, and everybody would just keep going and drink. They'd come back and nobody said anything. It was like the mafia. Or what did we used to say—the wet canteen at Creswell was like a handicapped home: that everybody in there was deaf, dumb and blind.'[69] Gleadhill is not sure whether a similar subculture existed among the officers or the officer cadets at Creswell.

It was more common for officers, in positions of authority, to protect each other. Brian McFarlane says that gay Army officers in the 1950s and 1960s used to meet at the Sportsman's Bar at the Australia Hotel in Melbourne and the Long Bar of the Australia Hotel in Sydney. Later in his career, from 1969 to 1975, McFarlane served at Victoria Barracks in Sydney (aside from a stint in Vietnam in 1971–72), just down the road from the emerging gay scene on Oxford Street in Darlinghurst. McFarlane used to frequent some of the gay bars, and even was known to take the odd straight soldier out to party. His bisexuality was, he speculates, an open secret that other officers tolerated because he was a good officer who was discreet. He explains: 'The Army did the right thing by me, despite perhaps general knowledge of my sexuality. And I think I did the right thing by the Army, but I didn't push the matter of my sexuality in their faces.'[70] McFarlane also speculates that when officers did face court martial for homosexuality, friendly officers would judge the defendant's quality as a soldier rather than care about his sexuality.

Of course, protection could also be dangerous when the crime was not just homosexual acts but sexual assault. The best example of this is Robert Percy, who reached the naval rank of commodore, was naval attaché at the Australian Embassy in Washington and was officer in command of the Western Australian forces in 1979 before his sudden retirement. Garry Gleadhill served as valet to the then Captain Percy from October 1971 to March 1972 on HMAS *Perth*. Gleadhill recalls that Percy would regularly entice drunk sailors, returning from nights out, to come to his cabin for another drink. Gleadhill remembers one sailor—a straight man—who was seeking a compassionate discharge and had a closed-door meeting in Percy's quarters. After some time Gleadhill opened the door and the stoker 'came screaming out of the shower, trying to pull—his shirt and everything hanging out—trying to pull his pants up and "Holy fucking hell!" he said, and pushed me, and bolted'. Immediately, Percy instructed Gleadhill to find the stoker: 'You get down there and tell that little mongrel if he says one word about anything his discharge is out of the window.' That stoker did get his compassionate discharge, and as far as Gleadhill knows he never reported what happened in Percy's cabin. Gleadhill was himself a victim of Percy's sexual predation. One time Percy ordered Gleadhill to 'undo your fly and show me where your appendix is'. On other occasions Percy tried to fondle Gleadhill when he was serving dinner, and another time Percy told Gleadhill to lather him up in the shower and tried to grope Gleadhill. The most traumatic incident is one that Gleadhill preferred not to describe. Shortly thereafter, Percy terminated Gleadhill's role as valet, using the excuse that he had prepared his eggs wrong, and Gleadhill was transferred off the *Perth*. Gleadhill discharged from the Navy in 1974.[71]

There were structural reasons why Gleadhill did not report Percy. Gleadhill's only option would be to report Percy to an Admiral's Requestmen, which was like an open court and happened three times a year. However, to lodge a report, Gleadhill would have to detail the traumatic experiences of sexual assault and needed the captain's signature to lodge the form. Decades later, Gleadhill went through the Defence Abuse Response Taskforce (DART) process, which concluded that he had indeed been abused by Percy, and was awarded a compensation package. What is more disturbing is that the DART case officer said to Gleadhill: 'This Captain Percy you were valet to ... We nearly had to get him his own filing cabinet ... I have never seen so many complaints and all these people come. That's what I work on.'[72] Percy is now dead, and allegations against him have been made public.[73] Yet for years this man continued to rise in the ranks of the Navy until his quiet discharge around 1979. It is hard to believe that other officers were unaware of Percy's predations, let alone his homosexuality. Indeed, DART's 2014 report noted: 'Many complainants expressed a strong belief that members of senior rank and Defence more generally were aware of the abuse, particularly within initial recruit and employment training establishments, and knowingly took no action to address or prevent it.'[74] Multiple reviews found that Defence culture discouraged reporting of abuse because complainants would face ostracism, and complaints against people of higher rank were even more dangerous because of the retribution a complainant might face.[75]

The number of servicemen who faced charges of indecent assault were few and far between. In the *Return of Naval Courts-Martial and Disciplinary Courts* for the entire period 1946–74, only six men were charged with indecent assault, two of whom were officers. Only two of these men were convicted, both other ranks. In November 1954 the press picked up a story about an Army corporal who pleaded guilty to the indecent assault of two National Service trainees. The complainants reported that the corporal 'had given him a drink at the camp, shown him indecent photographs and then interfered with him'.[76] Of course, it is quite possible that officers were charged and convicted for sexual assaults in other instances. But the available examples suggest that Defence authorities were less prone to target, charge or convict officers for crimes relating to homosexuality.

Policing homosexuality

Servicemen did not spend their entire lives cooped up on bases or RAN ships. They regularly were living, socialising and often working in cities and regional areas, mixing with civilians. As such, gay and bisexual men were sometimes participating in civilian gay subcultures, often preferring this to sex with other servicemen. Bart Hosking recalls of his time in the Citizen Military Forces from 1956: 'My only sexual encounter I had in the Army was with my driver and we were camped out in Tin Can Bay. Back then you had to be very careful.'[77] Hosking, like others, was more often visiting beats to cruise for sex with other men. Corporal Alan Lewis, who served in the Army from 1966 to 1972, recalls being picked up by men at a park not far from Woodside Barracks in Adelaide. When serving in Melbourne, Lewis would also visit the Australia Hotel to meet men and go back to their places.[78] David Collyer served in Townsville in 1969 with 2nd Battalion, Royal Australian Regiment, and he recalls his Army uniform attracting many men at beats:

> It was just so easy. No one back at Lavarack ever knew that one of my pairs of trousers had the bottom neatly cut out of the pockets. This made things very simple for someone I fancied to simply slip his hand in my pocket to sample what was on offer. Then it was a very short walk either down onto the beach or in among the aerial roots of the banyan trees to get down to real business. Others were always welcome to join in too if they wished. I distinctly remember one huge banyan tree, which could accommodate about 20 blokes hidden among its mighty root system. The heady smell of frangipanis will take me back there every time.
>
> If I could be bothered walking a couple of miles westwards around behind Castle Hill and on up towards Palarenda, there was many a fabulous orgy in which to participate. Eventually, feeling like Cinderella, I'd have to rush off back to Lavarack Barracks to check on all the straight blokes.[79]

Visiting beats was dangerous business, as men could be physically assaulted or be targeted by police. Newspapers from the 1940s to the 1960s often listed men charged for acts of gross indecency in the court reports. These reports sometimes mentioned when someone was a Defence member or veteran. Some examples include:

> A medical practitioner, an Army captain and a male shop employee whose names had been previously suppressed from publication until further notice, denied yesterday a joint charge that on the evening of New Year's Day, they wilfully committed an act of gross indecency with each other in a public urinal at the rear of Tattersall's Hotel, Hindley Street, Adelaide. (*Advertiser*, Adelaide, 7 February 1946)[80]

> A Regular Army warrant officer was ordered six months' imprisonment today for having attempted to procure an act of gross indecency … [He] enlisted in February, 1942, and while serving in Japan was in charge of Occupation Force farms. (*News*, Adelaide, 28 August 1951)[81]

> On Thursday, Austin Thomas Burns, 23, shop assistant, and George Joseph Anderson, 23, of the RAAF, were remanded until March 31 on charges of having committed acts of gross indecency in Ipswich on February 17. (*Courier-Mail*, Brisbane, 27 March 1954)[82]

> A young soldier was remanded for eight days in Perth Police Court today when he was charged with sodomy. Stanley Robert Bandle, 22, of Army Camp Swanbourne, is charged with committing the offence at Scarborough on May 20. He did not ask for bail. (*Mirror*, Perth, 2 July 1955)[83]

Ex-sailor James Timmins recalls that another sailor named Wayne was once arrested for trying to pick up men at the toilets at Sydney's Town Hall Station, and Wayne was discharged.[84] These public transgressions thus represented another form of exposure that violated unspoken norms about discretion.

Melbourne's *Truth* newspaper took a particular interest in covering homosexual scandals in the services. One headline from June 1956 that captured the paper's interest in the topic read 'Vice Shock in Army Camp'. This article reported on the discovery of 'an unsavoury cell of homosexuals' at the National Service training camp in Puckapunyal, Vic, and the discharge of five soldiers.[85] In October 1951 *Truth* reported on a flight sergeant, the most senior tailor in the RAAF, who was accused of indecently groping and then propositioning an aircraftman during a trousers fitting.[86] In one of the few known cases that did involve an officer, *Truth* in 1950 reported on a flight lieutenant test pilot arrested and court martialled for homosexual relations with a young recruit. They were known to spend time together, with the pilot even sometimes taking the recruit for flights. Then the young recruit went AWL while on guard duty one evening. They found him the next morning leaving the officers' mess, smelling of liquor. It was when RAAF Police questioned the recruit about why he went AWL that he confessed to a homosexual relationship with the pilot. RAAF Police shortly thereafter detained the pilot and charged him with two counts of disgraceful conduct of an indecent kind.[87]

The recruit was court martialled and sentenced to thirty-five days' detention; a week later the pilot, too, was court martialled. This entire case reinforces many of the patterns of how the forces treated gay servicemen in the period from the 1950s to the 1970s. First, the police were not searching for homosexuals, but rather the case came to their attention only because of another investigation. Because the two men were indiscreet—and especially because the press got wind of the case—the RAAF could not turn a blind eye and court martialled the officer. Importantly, in a foreshadowing of practices prevalent after 1974, the pilot testified to some unseemly RAAF Police tactics. They questioned him for hours, despite the fact that he was fatigued from having flown for two days and was under significant stress. He was given the impression that if he admitted to the homosexual acts, he would be able to avoid a court martial and publicity. RAAF Police denied ever implying that the pilot could avoid a court martial. Perhaps the pilot misread the situation; perhaps the RAAF Police did make that misleading proposition; or perhaps they did make that suggestion and meant it at the time, but because of the publicity in *Truth* they had to prosecute the pilot. Regardless, the outcome of the court martial was that the pilot, a decorated war hero, was dismissed from the RAAF.[88]

The pilot's case was exceptional not only because it was reported in the press but also because it went to a court martial instead of being dealt with administratively. One RAAF investigation in 1954 commenced in Brisbane but spread to Darwin when the suspects confessed to having performed homosexual acts with other airmen. RAAF Police investigators recovered intimate letters 'written in a strain bearing strong leanings and suggestions to homo-sexuality' and nude photographs. They then elicited confessions that detailed their sexual acts, including mutual masturbation, oral sex and anal sex. The investigation spread like a net as the confessions named other airmen and a sailor. Those airmen who signed confessions were discharged SNLR; the provost marshal advised that there were not sufficient grounds to discharge the two who would not sign confessions but suggested they be pressured to request discharges. The investigation snagged one officer, whose commission was cancelled, and the provost marshal wrote: 'Our present policy is not to charge unless there is an officer and an airman involved ... As you are no doubt aware, we try to prevent any publicity.'[89]

Finding data on how many people were investigated, prosecuted and dismissed in the three services is difficult because the information is kept in police files. A researcher would need to comb every file to identify which investigations were for homosexuality, but privacy provisions in the *Archives Act*—which would require that such files be redacted before they could be examined—preclude this option. There are a few records, however, that hint at the number of investigations and discharges. A letter attached to the above-mentioned RAAF investigation in 1954 said that 'I suppose [this year] to date we have discharged about 25'.[90] None of these cases went to court martial, which makes it even harder to trace them. RAAF Statements of Courts Martial are available for the 9.5 years from January 1954 through June 1964. During that period, only four men (two of whom were British RAF members), all from other ranks, faced charges for buggery, gross indecency or disgraceful conduct of an indecent kind; all were found guilty, and three of the men were discharged.[91]

Major-General Cedric Maudsley Ingram Pearson testified before Western Australia's Honorary Royal Commission Appointed to Inquire into and Report upon Matters Relating to Homosexuality in 1974. He said that there had been forty-four investigations in the Army during the five-year period 1969–74, uncovering twenty-one confirmed cases of homosexual behaviour. Statistics were not available for the RAAF or RAN, although the Navy estimated approximately eight per year.[92] There is no more information about these forty-four investigations, such as a gender or rank breakdown.

The other available source is the Navy's semi-annual *Return of Naval Courts-Martial and Disciplinary Courts*. These documents contain information about all courts martial in the Navy, including those charged for sodomy, acts of gross indecency, or indecent assault. Between 1946 and 1970, fourteen men faced courts martial for these charges, only two of them officers. Only three other ranks were found guilty, and even one of these convictions was quashed by the Naval Board.[93] Perusing these fourteen court martial files, a common thread is that almost all of them relied solely on eyewitness testimony. This meant that, when the sailors denied the charges, there was reasonable doubt to find them not guilty. The guilty cases had more evidence. The most dramatic example was one where prosecutors tabled evidence taken from the accused's locker, which included a series of erotic stories he had written, and a drawing of two men masturbating each other.[94]

Finally, there were those men who did not need any disciplinary hearings because they voluntarily confessed to being gay to discharge. It was the job of divisional officers like John Dunlop to interview those men and confirm whether they were actually gay. Dunlop reckons that probably only two of the men he ever interviewed were gay, and he permitted them to discharge.[95] Importantly, there is no suggestion that self-confessed homosexuals required any of the physical evidence dictated in the policy nor any additional police investigation.

Conclusion

As this chapter shows, there was a surprisingly tolerant attitude towards male homosexuality in the period of the 1950s to 1970s, but only so long as homosexual acts were conducted within certain bounds. Indeed, Hadrian wrote in 1972: 'From my own observations I would say that the troops themselves are fairly tolerant, and have, in the main, less prejudice than the general population.'[96] Gay/bisexual men needed to be discreet, knowing that exposure—whether it be via being caught in the act, exposed through other police investigations, reported by hostile colleagues or caught by civilian police—could lead to discharge. Gay and bisexual men were often adept at operating within these unspoken boundaries, finding sexual opportunities on ships, at known beats, with other gay/bisexual men and even with straight men who experimented with situational homosexuality.

Throughout the oral histories and archival records of this era is an important common thread about service approaches to homosexuality: service police were not actively hunting gay or bisexual men. But there was a subtle shift taking place in the early 1970s, illustrated well by the story of James Timmins. He had been serving in the Navy since 1967 and by the early 1970s was regularly visiting gay bars in Sydney. He was even in a long-term relationship with an ex-soldier. Then in 1973, naval police hauled Timmins in for an interview and threw questions at him, accusing him of being a homosexual. Timmins is confident that he came to police attention when another gay sailor was caught and named others in the hopes of saving his boyfriend's career. Timmins recalls the naval police playing good cop/bad cop and being really determined for him to name other gay men.[97] As chapter 2 shows, the pressure to name others through severe interrogations was already common practice in the women's services, and, as chapter 3 explains, this would become the mainstay for investigating LGB people throughout the Australian Defence Force from 1974.

Notes

1 Smaal, *Sex, Soldiers and the South Pacific*, pp. 99–100.
2 Peter Stanley, *Bad Characters: Sex, Crime, Mutiny, Murder and the Australian Imperial Force* (Millers Point, NSW: Pier 9, 2010), pp. 141–4.
3 Australian War Memorial: AWM15, 18197.
4 'A horrid hero—soldier sergeant severely sentenced', *Truth* (Melbourne), 13 March 1915, p. 6; '"An atrocious villain"—Todd's terrible turpitude', *Truth* (Melbourne), 25 March 1916, p. 3.
5 Smaal, *Sex, Soldiers and the South Pacific*, p. 114.
6 Garry Wotherspoon, *Gay Sydney: A History* (Sydney: NewSouth, 2016), pp. 70–90; Smaal, *Sex, Soldiers and the South Pacific*, pp. 15–65.
7 The Department of the Navy regularly published *Return of Naval Courts-Martial and Disciplinary Courts*, listing all courts martial for a year or a few years. All these documents for the period 1916–46 are available from National Archives of Australia (hereafter NAA), MP151/1, 445/201/123.
8 Department of Defence Minute Paper, 'Unnatural Offences', 14 February 1939, in NAA A1813, 321/251/1.
9 Willett and Smaal, '"A homosexual institution"', pp. 23–40; Smaal and Willett, 'Eliminate the "females"', pp. 233–50.
10 Smaal, *Sex, Soldiers and the South Pacific*, p. 107.
11 David Ritchie, interview with Noah Riseman, 7 July 2015, Sydney; Graham Willett, Wayne Murdoch and Daniel Marshall (eds), *Secret Histories of Queer Melbourne*, reprint edn (Parkville, Vic: Australian Lesbian and Gay Archives, 2017), pp. 42–6.
12 Graham Willett, 'The darkest decade: Homophobia in 1950s Australia', in *The Forgotten Fifties: Aspects of Australian Society and Culture in the 1950s*, ed. John Murphy and Judith Smart (Melbourne: Melbourne University Press, Australian Historical Studies, 1997), pp. 120–32; Robert Reynolds and Shirleene Robinson, *Gay and Lesbian, Then and Now: Australian Stories from a Social Revolution* (Carlton, Vic: Black Inc., 2016), pp. 7–37; Graham Willett, *Living Out Loud: A History of Gay and Lesbian Activism in Australia* (St Leonards, NSW: Allen & Unwin, 2000), pp. 3–18; Wotherspoon, *Gay Sydney*, pp. 107–39.
13 Graham Jamieson, interview with Noah Riseman, 7 July 2017, Berkeley Vale, NSW; Thomas Goldsby service record, 311222.
14 Hadrian, 'Homosexuality in the forces', p. 4.
15 Bob Flanagan, interview with Noah Riseman, 24 April 2018, Cessnock, NSW.
16 Bob Flanagan, interview.
17 Brian Allen, interview with Noah Riseman, 19 November 2018, Melbourne.
18 Hadrian, 'Homosexuality in the forces', p. 4.
19 Trevor Robinson, interview with Noah Riseman, 18 July 2015, Brisbane. See also National Library of Australia, Trevor Robinson, interview with Clive Moore, Australian Lesbian and Gay Life Stories Oral History Project, 10 October 2013, ORAL TRC 6510/4; David Hardy, *Bold: Stories from Older Lesbian, Gay, Bisexual, Transgender and Intersex People* (Panton Hill, Vic: Rag and Bone Man Press, 2015), pp. 184–9.
20 Tony Whelan, interview with Noah Riseman, 2 May 2017, Melbourne.
21 David Morrison, interview with Noah Riseman, 18 November 2015, Melbourne.
22 Brian McFarlane, interview with Noah Riseman, 10 July 2015, Bowral, NSW. On Brian McFarlane's life and career, see also Riseman, Robinson and Willett, *Serving in Silence?*, pp. 17–37; Brian McFarlane, *We Band of Brothers: A True Australian Adventure Story* (Newcastle, NSW: Exact Print + Design, 2000).

23 Gavin Hart, 'Sexual behaviour in a war environment', *Journal of Sex Research* 11, no. 3 (1975): 225.
24 Paul Ham, *Vietnam: The Australian War* (Pymble, NSW: HarperCollins, 2007), pp. 278–88; Mark Dapin, *The Nashos' War: Australian National Servicemen and Vietnam* (Melbourne, Vic: Viking, 2014), pp. 199–205. Dapin's historical novel *R&R* (Scoresby, Vic: Penguin Books, 2015) is specifically about Australians in Vung Tau.
25 Brian Allen, interview. There are also notes in Allen's service record commenting on his work with the local Vietnamese community in Vung Tau. See Brian Allen service record, 311259.
26 Bradford, *The Gunners' Doctor.*
27 Alan Lewis, interview with Noah Riseman, 20 July 2015, Sunshine Coast.
28 'Peter', interview with Noah Riseman, 28 May 2017, Coffs Harbour.
29 Sebastian Junger, *Tribe: On Homecoming and Belonging* (London: HarperCollins, 2016).
30 David Bradford, interview with Noah Riseman, 29 July 2015, Melbourne; Bradford, *Tell Me I'm Okay*, p. 41.
31 'Peter', interview.
32 J.G. Pairman, 'In which we serve', *OutRage* (December 1988), p. 39.
33 Alan Lewis, interview.
34 Collyer, *From Fishnets to Foxholes and Back Again*, p. 47.
35 David Bradford, interview. More on Bradford's time in Vietnam is available from Bradford, *Tell Me I'm Okay*, pp. 33–53, and *The Gunners' Doctor.*
36 Brian McFarlane, interview.
37 Lorenzo Montesini, interview with Noah Riseman, 2 March 2016, Sydney. See also Montesini, *My Life and Other Misdemeanours*, p. 165.
38 Shilts, *Conduct Unbecoming*, p. 150.
39 Leon Fry, interview with Noah Riseman, 5 March 2016, Sydney.
40 Pairman, 'In which we serve', p. 39.
41 Paul R. Deslandes, 'Situational homosexuality', in *Global Encyclopedia of Lesbian, Gay, Bisexual, Transgender, and Queer (LGBTQ) History*, ed. Howard Chiang, Anjali Arondekar, Marc Epprecht, Jennifer Evans, Ross G. Forman, Hanadi Al-Samman, Emily Skidmore and Zeb Tortorici (Farmington Hills, MI: Charles Scribner's Sons, 2019), pp. 1501–7.
42 Shilts, *Conduct Unbecoming*, p. 44.
43 Brian McFarlane, interview.
44 Smaal, *Sex, Soldiers and the South Pacific*, p. 73.
45 'C', testimony to Western Australia, *Honorary Royal Commission Appointed to Inquire into and Report upon Matters Relating to Homosexuality* (hereafter *Honorary Royal Commission*), transcript vol. 1, 2 May 1974, Western Australia State Records Office, AU WA S1989—cons5527 1, p. 41.
46 Pairman, 'In which we serve', p. 39.
47 Suzanne McDonnel and Peter Lalor, 'Ousted army gay is still bitter', *Sunday Age*, 21 June 1992, p. 2.
48 Alastair Cooper, '1955–1972: The era of Forward Defence', in *The Royal Australian Navy: A History*, ed. David Stephens (South Melbourne: Oxford University Press, 2001), pp. 203–8.
49 Wally Cowin, interview with Graham Willett, 13 September 2016, Brisbane. See also Riseman, Robinson and Willett, *Serving in Silence?*, pp. 85–6.
50 Name redacted, statement form, 9 February 1966, NAA A1813, 351/250/275.
51 See Tom Frame, *Where Fate Calls: The HMAS Voyager Tragedy* (Sydney: Hodder & Stoughton, 1992); Cooper, '1955–1972', pp. 201–3.
52 'Xavier', interview with Noah Riseman, 10 August 2016, Melbourne.

53 'Xavier', interview.
54 Bérubé, *Coming Out Under Fire*, pp. 57–66.
55 John Dunlop, interview with Noah Riseman, 22 May 2017, Melbourne.
56 'Xavier', interview.
57 'Confidential Australian Navy Order 35/66', 6 July 1966, available from Australian Lesbian and Gay Archives (hereafter ALGA). This and earlier versions of the document since 1954 also available from NAA A1813, 321/251/1; NAA A1813, 321/201/39.
58 'Peter', interview.
59 'Confidential Australian Navy Order 35/66', 6 July 1966, available from ALGA; NAA A1813, 321/251/1.
60 The archival reference for this and other court martial files are not provided here because they identify the individuals charged. These men might still be alive and might not want this information disclosed. Please contact the authors for further information if required.
61 'Confidential Australian Navy Order 1–2/69', 21 February 1969, available from ALGA.
62 Wally Cowin, interview. See also Riseman, Robinson and Willett, *Serving in Silence?*, p. 88.
63 Wally Cowin, interview; Riseman, Robinson and Willett, *Serving in Silence?*, pp. 88–92.
64 Second Naval Member, 17 March 1966, NAA A1813, 351/250/275.
65 NAA A1813, 351/250/275.
66 John McGeorge, 'Outpatient record', name redacted, 15 February 1966, NAA A1813, 351/250/275.
67 'Report of Investigation into Homosexual Activities in HMAS *Melbourne*', 1966, NAA A1813, 351/250/275.
68 Richard Williams, interview with Noah Riseman, 29 November 2018, Griffith, NSW.
69 Garry Gleadhill, interview with Noah Riseman, 20 April 2017, Branxholm, Tas.
70 Brian McFarlane, interview. See also Riseman, Robinson and Willett, *Serving in Silence?*, pp. 30–3.
71 Garry Gleadhill, interview.
72 Garry Gleadhill, interview. See also Gleadhill DART submission, courtesy Garry Gleadhill.
73 Joseph Catanzaro, 'Navy chief raped us: sailors', *West Australian*, 16 June 2012, https://thewest.com.au/news/australia/navy-chief-raped-us-sailors-ng-ya-311453, retrieved 27 June 2017.
74 Defence Abuse Response Taskforce, *Report on Abuse in Defence* (Canberra: Defence Abuse Response Taskforce, 2014), p. 82.
75 Gary Rumble, Melanie McKean and Dennis Pearce, *Report of the Review of Allegations of Sexual and Other Abuse in Defence: Facing the Problems of the Past*, vol. 1 (Canberra: Commonwealth of Australia, 2011), pp. 21–4; Defence Abuse Response Taskforce, *Report on Abuse in Defence*, pp. 81–4.
76 'Army corporal for trial', *Morning Bulletin* (Rockhampton, Qld), 5 November 1954, p. 4; *Central Queensland Herald* (Rockhampton, Qld), 11 November 1954, p. 10; 'Committed on camp indecency', *Maryborough Chronicle* (Maryborough, Qld), 5 November 1954, p. 1.
77 Bart Hosking, in Hardy, *Bold*, p. 193.
78 Alan Lewis, interview.
79 Collyer, *From Fishnets to Foxholes and Back Again*, pp. 9–10.

80 'Adelaide police', *Advertiser* (Adelaide), 7 February 1946, p. 9. See also 'Adelaide police', *Advertiser* (Adelaide), 2 March 1946, p. 11. This case was later dismissed; see 'Criminal sittings', *Advertiser* (Adelaide), 13 March 1946, p. 12.
81 'Soldier gaoled for 6 months', *News* (Adelaide), 28 August 1951, p. 7. See also 'Supreme before Mr Acting Justice Paine', *Advertiser* (Adelaide), 29 August 1951, p. 14.
82 'Three in court; indecency alleged', *Courier-Mail* (Brisbane), 27 March 1954, p. 12.
83 'Soldier charged', *Mirror* (Perth), 2 July 1955, p. 19.
84 James Timmins, interview with Shirleene Robinson, 17 August 2016, Sydney.
85 'Vice shock in army camp', *Truth* (Melbourne), 23 June 1956, p. 1.
86 'RAAF clips wings of senior F/SGT tailor', *Truth* (Melbourne), 27 October 1951, p. 10.
87 'RAAF mute as scandal hits service and parents', *Truth* (Melbourne), 4 March 1950, p. 1; 'RAAF court-martial hears youth's alleged story of indecent conduct', *Truth* (Melbourne), 18 March 1950, p. 3; 'RAAF ace dismissed from service for disgraceful affair with AC1', *Truth* (Melbourne), 23 March 1950, pp. 1, 5.
88 'RAAF ace dismissed from service for disgraceful affair with AC1', *Truth* (Melbourne), 23 March 1950, pp. 1, 5. We have not included the reference to the pilot and recruit's courts martial because that would out them, and they might still be alive. Please contact the authors for further information.
89 NAA A161, 705/7/P1.
90 NAA A161, 705/7/P1.
91 NAA A9177, 703/16/P1 Part 2; NAA C2439, 703/1/P1 Part 2.
92 Western Australia, *Honorary Royal Commission Appointed to Inquire into and Report upon Matters Relating to Homosexuality*, Transcript Vol. 3, 25 June 1974, pp. 596–7.
93 NAA A1813, 321/10/127; NAA MP151/1, 455/201/234; NAA K1259, A/1993/04651; NAA A1814, 321/10/127; NAA A463, 1958/86.
94 We have not included the reference here because that would out this member, who might still be alive. Please contact the authors for further information.
95 John Dunlop, interview.
96 Hadrian, 'Homosexuality in the forces', p. 4.
97 James Timmins, interview.

Chapter 2

The women's services, 1950–85

In the 1960s, 'Helen' enlisted in the Women's Royal Australian Army Corps (WRAAC) because 'I wanted to meet other lesbians. Heaps of lesbians were in the Army in the '60s but you had to hide, oh God yes.' Helen found a career and a social world that she enjoyed. After three years of service, though, Helen was dishonourably discharged when her sexuality was discovered. She remembers that she had to report to a 'hypocritical woman major, who I knew was a lesbian herself, and she gave me a pen and pad and told me to write down all the names of others in the Army who were "like myself"'. Helen 'refused to dob anyone in' and was forced out of the career she had liked.[1] Helen's story is far from unusual and captures many elements of the lesbian and bisexual experience in the women's services.

From the 1950s through to the mid-1980s, the women's services provided a space where a vibrant yet hidden lesbian subculture thrived. Many women discovered and explored their sexuality in this space. Aware of this from very early on, and guided by the medical model of homosexuality prevalent at this time, the services embarked on aggressive so-called witch-hunts, designed to police women's sexuality and remove non-heterosexual women from the military.[2] Women who were found to have engaged in same-sex relationships or conduct were forced out, often after intimidating interrogations that could cause long-lasting psychological trauma.[3] As the military wanted to prevent the general public from believing that the services were encouraging lesbianism, authorities tried to keep cases where women were forced out because of their sexuality very quiet throughout the 1970s and 1980s.[4] Official paperwork rarely mentioned the real reasons women were discharged. However, the aggressive treatment meted out to lesbian and bisexual women from the 1950s onwards formed the prototype that was widened to persecute gay and bisexual men after 1974.

Women have a long history of serving in Australia's military. This service took on particular importance during World War II. In 1944, almost 50 000 women served in female-only branches of the military, and thousands more were engaged in the civilian Australian Women's Land Army.[5] Although the women's branches of service were disbanded after the end of World War II, they were revived in the 1950s in the shadow of growing international conflicts to release male labour for combat roles. Women always worked alongside men, but were enlisted through the separate, women-only services. The Women's Royal Australian Air Force (WRAAF) operated until 1977, and the Women's Royal Australian Naval Service (WRANS) and Women's Royal Australian Army Corps (WRAAC) both operated until 1985. By this time Australian women from all three services were integrated into the broader Australian Defence Force, serving in the same units as men.[6]

During the 1960s, the Australian Government undertook actions that increased the tactical significance of the women's services. In November 1964, the Menzies government reintroduced National Service for men. In 1965, it expanded the commitment of Australian troops to Vietnam. Some 60 000 Australians served in this war with a peak strength of more than 8300 personnel in mid-1969.[7] There was a mounting need for women who could augment the competencies of the forces at a time when servicemen were deployed overseas.

Official anxieties

Although lesbianism was never illegal in Australia, it appears that the Australian armed services were far more concerned about homosexuality between women than homosexuality between men for much of the post-war era. As chapter 1 argued, the 1950s and 1960s were a time of silence around male homosexuality, which meant that service police gave little attention to the presence of gay or bisexual servicemen. Civilian society also maintained a silence around lesbianism, yet the military world was far from silent about the topic. There is a substantial history of twentieth-century discourse suggesting that military services had the potential to 'masculinise' or awaken the sexuality of women—and not just the heterosexual sexuality of women.

During World War II, when female branches of the Australian military were first formed, popular periodicals published material depicting military women as mannish or in danger of losing their femininity and forming same-sex bonds that overrode attachments between men and women. Historian Ruth Ford has argued that during this era, 'It was imagined that the militarisation of women would produce lesbian bodies, subjectivities and relationships through the uniform, martial bearing and same-sex environments necessitated by military training.'[8] The military itself attempted to counter this image by producing promotional material and public commentary that emphasised the femininity of servicewomen and their innate heterosexuality. In 1941, for example, when she was appointed assistant controller of the Australian Women's Army Service (AWAS) in New South Wales, Eleanor Manning emphasised: 'Women who join the AWAS will be given every encouragement to remain entirely feminine. Women's service that tends towards the masculine is just silly.'[9]

When the women's services were reconstituted from 1950, the official anxieties over women's sexuality and their transgression of gender roles, which had been evident during World War II, were also reanimated. As Ford has argued, 'There was much more anxiety and action over lesbianism in the armed forces than in the wider community, where lesbian existence had usually been private or secret.'[10] Ford attributes this increased anxiety to a number of factors. Doctors and sexologists of this era had maintained that there was a link between lesbianism and women wanting to serve in the military. The presence of women who were perceived to have masculine characteristics in the service only served to increase this apprehension. The all-female environment was further believed to encourage lesbianism.[11] While men who served in the Australian military were seen to exemplify the highly desirable attributes of masculinity, dominance and nationalism, women who signed up were challenging the narrow concepts of femininity they were meant to embody and transcending the expectations of how women should behave and present. As Janette Bomford has argued, 'The entry of women into the armed services is a story of women's negotiation for a place within the most traditionally male profession.'[12]

In the 1950s, the US military also noted that sexuality or identification with masculine roles might have motivated some women to join the US Women's Army Corps (WAC). In their study of a lesbian community in Buffalo, New York, in this era, Elizabeth Lapovsky Kennedy and Madeline D. Davis noted the centrality of military service for many lesbian women.[13] In 1953, as a result of a belief that lesbian or masculine women were joining the WACS, the US Army passed new regulations, which advised the psychiatric officers who examined women before they were accepted in the WACS that 'particular attention must be given to the nature and quality' of the motives guiding women who were volunteering to serve. The regulation advised that 'women may be expected to volunteer for service for many reasons: patriotism, financial and personal security, masculine identification, refusal to accept female role, escape from environmental or situation conflicts, occupational change, glamor and excitement, opportunity to express femininity etc'. The regulation further advised, 'These motives cannot in themselves be considered acceptable or non-acceptable. They require exploration to reveal how transitory, compulsive, or immature they may be in stimulating possibly ill-considered action.'[14] Numerous Australian newspapers widely publicised the Americans' adoption of this new regulation.[15] This psychiatric treatment of 'masculine identification' and a 'refusal to accept [a] female role' reflects the medical model of female homosexuality that was also prevalent in Australia at this time.[16]

From the 1950s through to the 1970s, female same-sex desire was constructed as 'unacceptable'—a construction that was reinforced through broader cultural silences on the topic.[17] The Australian military reflected this construction of lesbianism as unacceptable, maintaining that the presence of homosexual or bisexual women would lower morale; that recruits would be susceptible to homosexual advances; and that gay and bisexual women would be vulnerable to blackmail if they were permitted to serve. As chapter 3 will show, these same arguments were consistently used to justify banning all LGB Defence members from 1974. Moreover, as previously noted, the medical model of homosexuality treated homosexual desire as a condition that should be treated as a psychological or pathological disorder to be 'corrected' in this period.[18]

Although same-sex conduct between women did not break any laws in the period between the 1950s and 1985, women who were believed to be 'untreatable' lesbians or bisexual were forced out of the services, usually as quietly as possible. Even women who were suspected of engaging in homosexual conduct came under aggressive scrutiny. Official discharges could be dishonourable and would sometimes note that the women had engaged in 'conduct prejudicial to the corps'.[19] Oral histories indicate that different discharges were issued throughout the services, with the WRAAF arguably less likely to issue dishonourable discharges. Ford has suggested that during the 1950s and 1960s, women of higher rank were sometimes able to avoid discharge and were instead transferred to different locations to deflect attention away from their sexuality.[20] Oral histories confirm that this continued to be the case for some women of higher rank during the 1970s and 1980s.

Media attention in 1973 on the treatment of lesbian and bisexual women stimulated the services to evaluate their processes for investigating and discharging all LGB members. Two women who had been discharged because of their sexuality requested that their cases be investigated further and even wrote to the Defence Minister. One of those women's service records makes no mention of sexuality in any documents relating to her discharge. Instead, she was investigated for fraud over a paid claim entitlement for a trip with another WRAAF (possibly the other discharged WRAAF), which she never took. This same WRAAF went to the mainstream media and to the gay and lesbian press. In an interview on *A Current Affair*, the WRAAF described a long, intimidating police interview in which the RAAF Police asked questions like 'what acts I have carried out with Jenny … very personal questions concerning our Lesbian behaviour; whether we had carried out embracing … I can't go into more detail … embracing and kissing and generally making love.'[21] Following the media attention, the RAAF briefed the minister with an explanation of the investigation and explained that, 'whilst homosexuality between consenting female adults is not a criminal offence anywhere in Australia, the Service finds such behaviour unacceptable on three [sic] grounds'. It maintained:

(a) there is a security risk through blackmail of participating persons;
(b) the Service has a responsibility to members of tender years and their parents to avoid exposing other members to association with or temptation in regard to homosexuality;
(c) there is the reputation of the Service to consider in particular any suggestion that it condones such practices would bring it into disrepute in wide sections of the community; and
(d) there are frequently problems of discipline involved.[22]

As chapter 3 outlines in more detail, the subsequent policy designed in 1974 was developed in conjunction with a growing social awareness about homosexuality that was starting to challenge the broader cultural silence on this topic.[23] In 1976 Sir James Killen, who was then the Minister for Defence, approved a statement which asserted that 'the policy reflects that, although homosexual behaviour is not a frequent occurrence in the services, it is not acceptable. It is, however, necessary to distinguish between different levels of behaviour.'[24] This statement drew a distinction between men and women whose sexuality was considered 'untreatable' and those who might be considered to be 'situational' lesbians or gays. This reflected views that had existed since the post-war women's services were founded in the 1950s.[25] In 1982, a representative of the Department of Defence indicated that official attitudes had not softened. The representative stated that 'homosexuality among Australian servicemen and women was bad for discipline and morale and would not be accepted under any circumstances'. Furthermore, 'If a person was a "known homosexual" he or she would be asked to leave the services—even if the activities took place in private.'[26]

The role of women in the services

From the beginning of their re-formation in the 1950s, the women's services emphasised the important role that women could play, even if it was largely framed as enabling more men to be deployed. In 1952, the then Assistant Director of the WRAAC in Tasmania, Lieutenant A.M. Jaggers, asserted: 'Women are needed for the home base organisation to release men for service in field units and are capable, as they proved during the Second World War, of filling many posts in the administrative sections of overseas forces. Outside the actual fighting roles, there is hardly any task in the army which women cannot perform and perform well.'[27]

In 1953, the Adelaide *Advertiser* praised the 'important part in the defence of Australia' women in the Women's Auxiliary Australian Air Force had played during World War II and emphasised the present 'equally important—if perhaps seemingly less urgent vital role for the women's component of the RAAF, now the Women's Royal [Australian] Air Force, in Australia's defence plans'.[28] Similarly, a 1955 article in the *Argus* newspaper said: 'No story of the Navy would be complete without a word of appreciation to the girls in blue of the Women's Royal Australian Naval Service ... The Navy, already heavily under-staffed, would be in a more serious plight if it could not call on girls to do the work of men.'[29] In the 1950s, when the women's services began operating, career choices for women in mainstream society were limited. In contrast, the women's services were able to position themselves as offering varied and exciting careers. It is estimated that about 10 500 women served in the WRAAC alone, and probably a similar number in the Army Reserve over a period of thirty-three years.[30]

Promotional material from the 1950s onwards not only emphasised the importance of women serving; it also drew attention to the various types of training and careers that could emerge from military service. In 1951, an Adelaide newspaper told readers about the stimulating occupational prospects available in the WRANS: 'New recruits will have general training, which will enable them to be transferred to suitable branches. These branches are communications, clerical, clothing, victualling stores, and messing. They may become wireless telegraphists, office workers, store assistants, cooks, or stewards. Mechanically minded girls with manual dexterity, such as touch typists, make good telegraphists.'[31] Perhaps unsurprisingly, there was significant interest from women who wished to join. Within the first two weeks of the WRANS recruitment opening, 1500 applications were received for the 250 billets that were available.[32]

In 1952 an advertisement focusing on the opening of a new Officer Cadet School for the WRAAC in Portsea, Vic, asserted: 'Young women will be expertly trained in regimental and staff duties, organisation, administration, personnel management, military law and current affairs. This is an exceptionally attractive career opportunity for applicants with character and capacity for leadership.' Furthermore, prospective applicants to the WRAAC were told that 'pay during training and service is excellent and gratuity accrues, payable on retirement'.[33] In 1954, the Adelaide *Advertiser* emphasised the opportunities that awaited women who signed up for the WRAAF, declaring:

> Members of the WRAAF receive the same privileges and benefits as the men of the RAAF. Their pay is liberal. They receive an initial free issue of uniform and service clothing and an allowance for the upkeep of their uniform. They have free board and lodging, medical and dental care. And although on enlistment they are given the opportunity to choose the category they work under they are encouraged to train for other musterings if they wish.[34]

Although servicewomen arguably did have more opportunities than women in many other careers outside the services, they were treated markedly differently from men who served. It was not until 1968 that retention of married women in the Australian armed services was approved. Even then, as Janette Bomford notes, this retention was known as 'permissive retention' because it was not automatic and was applicable only if 'the Service has a need for her special talents, acquired through previous training and experience in the Service'.[35] Furthermore, the permission of the servicewoman's husband was also required. It was not until 1979 that equal pay for women in the services was introduced.[36] Women who served in the WRANS were not able to serve aboard ships until 1983.

Although there were certainly limitations on the roles women could play in the military in this period, the women's services still offered the opportunity to play important roles in Australia's military security. Noah Riseman and Richard Trembath have noted that in the post-war period, women undertook critical functions with strategic importance in the WRAAC, WRANS and WRAAF. One example was that women in all three services worked in the respective Signal Corps where 'tasks included enciphering and deciphering messages, operating several pieces of wireless equipment and working around the clock with access to top-secret information'.[37] Over time, the scope of work that women were able to undertake widened. In 1967, in a major step that recognised the important role they were playing during the Cold War, the first women from the WRAAC were deployed overseas to Singapore for twelve months service with 121 Signals Squadron.[38] During the 1970s and early 1980s, the services began to integrate their training, commenced direct enlistment into employer corps and increased the number of occupations open to women. It became inevitable that the single-sex services would be phased out completely, which they were by 1985.[39]

The unacceptability of homosexuality in the women's services

While heterosexual women were actively encouraged to join the military, homosexuality and homosexual women were framed as unacceptable. The issue of lesbianism was generally not discussed publicly by military officials before the 1970s, but when the topic was raised—particularly during education courses as part of recruit training—officials were clear that it was objectionable. Nola Strawbridge, who joined the WRAAC in 1962 and worked in the Survey Corps at Bendigo, remembers that homosexuality 'was never, certainly never stated, and never spoken about. But it just went without saying that you kept it under wraps. It wasn't something that we talked about, unless you were talking to someone who was the same.'[40] Carole Popham, who was sworn into the WRAAF in 1963, remembers receiving a lecture as a rookie on sexuality, in which the servicewomen were warned 'not to have anything to do with anybody who's gay. That was not on. In fact you have to report them because it's tut, tut, tut.'[41] Julie Hendy, who joined the WRAAC in 1964, remembers absorbing the message that homosexuality was against WRAAC policy: 'You certainly knew. There was no question about knowing that you'd be out on your ear.'[42]

It is also evident that the three women's services pre-screened women for 'homosexual tendencies' as part of the enlistment process. In 1966 Channel Seven's documentary television program *Seven Days* broke the cultural silence around lesbianism when it interviewed a woman who tried to sign up for the WRAAC but had not been accepted when her initial psychiatric interview revealed she was a lesbian. She told interviewer Anne Deveson, 'All of these questions I realise now were to trick me and find just exactly what I was, whether I was a normal girl as they put it in square terms, or if I was a homosexual.' She continued, 'It was a tremendous blow to me because this was the first time in my life there was ever anything that because of my homosexuality I wasn't going to be able to do.'[43]

Some servicewomen remember negotiating anti-homosexual policies. Elaine Crump, who joined the WRANS in 1976, remembers somewhat irreverently discussing the WRANS policy towards homosexuality. Somewhere in her recruit literature: 'There was something about homosexuality not being allowed and I remember asking about it and looking back at that now, I think, "Oh geez that was just so audacious of me." I remember asking, "Well, what happens if someone's found to be gay?"' She was told, 'You'd have to be discharged.' Ironically, Crump later found out that the woman who told her that homosexuals who were discovered would be discharged was herself a lesbian. This woman later told Crump that when this discussion had happened, 'She didn't know what to say.'[44]

Oral histories from some women indicate that officers were not uniformly hostile in policing homosexuality. Some women remember discreet toleration on the part of some superior officers and from heterosexual servicewomen. Lyndal Green recollects that 'there were so many of us' gay women in her WRAAC transport unit in Sydney in the early 1980s, and her male commanding officer

> loved the fact that there were all these gay women in his unit, because we didn't come to work complaining about our kids, we didn't have to take time off because our husband was sick. We didn't complain when we had our period. We just came in and got on with the job. If you wanted someone to go interstate we had no commitment per se, and we just did it.[45]

Overall, though, lesbian and bisexual women learned to hide their sexuality on the expectation that officials and superior officers would be hostile towards them.

Hiding their sexuality and relationships could be challenging. Servicewomen had to remain constantly on guard and be careful in their use of pronouns. They had to be very discreet about their activities and develop ways of explaining out-of-work social lives to heterosexual colleagues and superiors. Ronni Spencer, who joined the WRANS in 1979, remembers the strain of hiding her first relationship with a civilian woman:

> I found it really hard because I wanted to share my joy with other people, but I wasn't able to. People would say they're getting married or they'd done this, this and this at the weekend, and been out with their boyfriend, and I just had to say, 'I just hung around and caught up with friends and blah, blah, blah.'[46]

Spencer's relationship with a civilian ended because of the strain of hiding the relationship. Relationships with other servicewomen were common, partly because of the opportunity to find each other, but also because there was a mutual understanding of the pressures of hiding the relationship.

Motivations to join the services

Despite the fact that the Australian military tried to discourage women who were not heterosexual from signing up, lesbian and bisexual women still were driven to enlist in significant numbers. There were a number of reasons for this, including the same patriotism and family connections that motivated heterosexual women. Furthermore, the women's services were still an attractive employment proposition for those who wanted to avoid the limited career options and the resolutely heterosexual domestic expectations of marriage and children that were so ubiquitous in this period.[47] As Julie Ustinoff has argued, 'Australian society in the 1960s dictated that women were the custodians of the nuclear family, responsible for the welfare of husband, child and home.'[48] The impact of the women's movement of the 1970s resulted in a number of incremental gains towards the fuller inclusion of women in the workforce, yet it was not until 1984 that Commonwealth Parliament passed the *Sex Discrimination Act*.[49] Even so, stereotypes about the nature of masculinity and femininity persisted, and expectations about the social role of women that emphasised child-bearing and household duties endured throughout the 1980s and beyond.

Joining the women's services offered women, including those who had yet to realise or come to terms with their sexuality, a means of meeting other women who shared their sexuality. If women were able to conceal their sexuality successfully for long enough, the fact that they were single without children could even offer an opportunity for considerable career advancement. Helen, who joined the WRAAC in the 1960s when she was seventeen, points out: 'The higher the rank, the more likely the woman was lesbian, because straights mostly married and had to leave.'[50] 'Pencil', who joined the WRAAC in 1971, worked out that after she had served three years, only fifteen out of the eighty women who had joined up at the same time as her were left because 'In those days if you married you had to get out of the Army'. Although this could advantage gay women in some ways, she also points out that 'you had to have a fighting spirit about you, I tell you'. Pencil herself endured two brutal witch-hunts and investigations but managed to stay in the ADF for twenty-five years.[51]

For Nola Strawbridge, who joined the WRAAC in 1962 at seventeen, service offered her independence and a female-centred environment. Importantly, it also offered financial support and vocational training that would serve her throughout her entire working life. She had been thrown out of her home by her father after he had become aware of her sexuality and after she had undergone psychiatric treatment, including the administration of bromide, meant to 'address' her sexuality. Strawbridge was selected to join the Survey Corps. There she 'learnt all the techniques of surveying, cartography and lithography, printing of maps'. She enjoyed the work: 'It was great. It was interesting and I thought it was useful.'[52] She emphasises the security that enlistment provided at a difficult time in her life: 'At seventeen, with not a cent to my name, and no support, no family support, yeah. It's quite phenomenal really, the way, you know, just taking one step and the whole world opens up really.'[53]

Sandra, who joined the WRANS in 1965 at the age of seventeen, went straight 'off the farm, out of the convent, into the Navy'. She believes her brother's service in the Navy might have motivated her. She emphasises that in her small town, joining a convent or 'teaching or nursing' were the only career options that would have been open to her. After joining the WRANS, she was given training as a radio operator, learning Morse code among other skills, before serving in Canberra and Darwin.[54]

For Susie Struth, who grew up in Warrnambool in Victoria and joined the WRAAC in 1971, the magazines her mother used to receive played a pivotal role. She remembers, when she was thirteen or fourteen, 'Mum used to get the *Women's Weekly* every weekend [and] there was an ad in there for women in the Army. There was one for the Navy, one for the Air Force and one for the Army and I just thought, wow that looks fantastic!' She continues:

> I didn't want to be a teacher. I didn't want to be a nurse. I didn't want to be a secretary. I didn't particularly want to work in a shop and I wasn't pretty enough to work in the chemist, which was—pretty much what was open to girls in the '70s from a country town. So that's why this whole world opened up to me when I saw this ad, and I just went, yeah, that's me.[55]

It was the possibility of a more interesting life in the women's services that attracted Delia Quigley, who joined the RAAF at the age of seventeen in 1978. She faced the same limited employment options as Susie Struth, having also grown up in the Warrnambool district. Leaving school after completing year 11, Quigley initially took up a job working at a bakery. She had an awareness of the option of military service as she had seen others from her town join up, and her father had served with the RAF in Burma during World War II. She remembers feeling discontented with the options available to her in her town and as a woman:

> Staying in the country area wasn't what appealed to me. Working in a factory or woollen mills or those sorts of things didn't appeal to me. And you know there were lots of young girls that were teenagers that were just sort of immediately falling into relation-

> ships and that was the man they were going to marry. And a lot of people were getting married and having babies and you know—whatever order it happened in—and that didn't appeal to me. That didn't sit with me.[56]

Although military service might have required lesbian and bisexual women to hide their sexuality, this was not markedly different from life outside the military, where exposure could also come at significant personal and professional risk.[57] Rebecca Jennings has pointed out that in civilian society during the post-war period from the 1950s to the early 1990s, violence was a continuous risk for lesbians:

> Younger lesbians in particular experienced violence from their families and peers as well as from a range of authority figures, including medical professionals and the police. Women's experiences of random acts of physical and sexual assault also indicates the extent to which violence was used to contain lesbian visibility in public spaces.[58]

In this sense, for many women, the armed forces did not appear to be any more homophobic than the civilian world. Furthermore, the military also offered opportunities—professional, social and intimate—that were less accessible in the civilian world.

Desire and a subculture

Despite official censure, it is clear that the services provided many women with an opportunity to realise or explore their sexuality and form relationships. It was possible to find a subculture that afforded opportunities to socialise. Julie Hendy, who began her time as a WRAAC recruit in 1964, was subsequently sent to Mount Martha, Vic, to undertake a six-month Operator Keyboard and Radio (OKR) signals course. It was here that she first became truly aware of her sexuality when she fell 'madly in love' with another woman whom she knew was a lesbian:

> I was mad about her, and you have to look at yourself and think about that. And I realised that's that, and I was very comfortable with that. And I hadn't, I wasn't going to question it. It was me. I knew also that the Army didn't approve, so you'd be in deep shit if anyone found out, so you had to keep it under wraps.

At Mount Martha, Hendy became aware that there was a network of lesbian women who were serving, although 'it was very complicated too because you had to suss it out'. Once 'you got to know someone in the circle, they would say, "There's so and so and there's so and so." And so you get to know who's around.'[59] Such information had to be kept very quiet. Ford has noted that there was a 'massive witch-hunt' of lesbians in the military between 1964 and 1965, just as Hendy was realising her sexuality at Mount Martha.[60]

A female-only environment proved the catalyst for many women to explore their sexuality. Jennifer Jefferies signed up for the WRAAC in 1979 while engaged to a boyfriend and with very little understanding of homosexuality. She remembers:

> Right through WRAAC School, I had no concept of gay and not-gay, none. But my attraction to these women and just having fun with them was just the best. Just the best on so many levels. And I guess it was all stirring in me, but I had no idea how it manifested in them or how it looked. And then went to driver training, and then that's when I kissed a girl for the first time, and the rest of it started to evolve.[61]

Lyndal Green remembers that service in the WRAAC provided opportunities for romantic liaisons:

> Didn't hang around with civilians. Civvics? I don't think so. No, it was—we were within the unit, women that were within your corps. The lass that I was with was a corporal and then got promoted to sergeant. We were all in the same barracks. We all snuck around at night into each other's bedrooms. Did the commando run in the morning to get back before anybody saw us.[62]

Delia Quigley became aware of networks of gay men and lesbian women within the services at the same time as she realised her sexuality. Quigley met a partner, who was serving in the WRANS. She remembers: 'There was a little group of a couple of Navy lesbians and there was a whole bunch of Army lesbians and there was a couple of Air Force lesbians and there was a big bunch, a big mixed bunch, of gay service guys that would go and party and go to … different venues.'[63] WRAAC member Chris Sinclair similarly describes a network of house parties and pub meet-ups in Canberra in the late 1970s and early 1980s, many organised by her gay RAAF mate Dennis Carney, where gay and lesbian service members would socialise.[64]

In the WRAAC in the 1970s, Lyndal Green found other lesbian and bisexual women through 'rumour, and innuendo' as well as 'certain catch-phrases'. If a new soldier joined the unit and Green suspected she was gay, she would discreetly 'name drop. If they know the names of well-known gay women then they're obviously gay.'[65] Gen Ford, a captain who commenced in the WRAAC in 1983 and served in the Army until 1988, finds it hard to explain the signals. She says women would exchange glances with a sense of: '"I like the look of you." [Laughs] And they're going, "Mm, I like the look of you too." And it's, yeah, it's the gaydar. And you go—you cannot explain how it works for the life of you.'[66] Elaine Crump, who was deployed to Melbourne in the WRANS in 1979, remembers one occasion when she subtly picked up the verbal cues that a woman she was talking to was gay, and the social relationships that subsequently followed. She states:

> After you've sort of talked for a while and you've used the, you know, 'them' and 'they' and, you know, no personal pronouns, it becomes apparent and eventually it was sort of disclosed and then once it happens you realise that it's a whole world out there, that the Navy's full of them. So you know she knew people and then we met up and then we'd go for camping weekends and we'd go into town, we'd go to gay pubs, and that's when it just all opened up for me. It was wonderful.[67]

Sports—including competitions between the women's services—also provided a popular means for women to connect with each other between the services. Nola Strawbridge remembers going to watch the woman who would become her first female partner in the WRAAC play basketball. Sandy Ashton, who joined the RAAF in 1979, formed her first relationship with a woman at the age of twenty-one through sports in the services: 'I was based in Adelaide at the Edinburgh base there, playing in my soccer team. We had a soccer team called the RAAF Rockettes.' Ashton remembers, 'And so it was during this time that I met this woman ... she was a PE instructor and we used to get on really well and became quite close friends.' While Ashton was still wrestling with internalised homophobia, 'that was the start of my first relationship', and it lasted for just under two years.[68] Similarly Jenny Robertson, who joined the RAAF in 1984, remembers, 'An excellent way to meet women was to play sport. Principally, the main sorts of sports that lesbians seemed to gravitate towards were softball, basketball and hockey.' Robertson met a partner she was with for thirteen years through inter-service softball.[69] Police investigators were also conscious that lesbian and bisexual women often played sports, so investigations often targeted sports teams.

Witch-hunts

Although it was possible to find a lesbian subculture within the women's services between the 1950s and 1980s, this subculture had to remain hidden from officials. This was frequently very difficult, as the military regularly embarked on so-called witch-hunts intended to unearth lesbians and bisexual servicewomen.[70] Women who were suspected of being lesbians or bisexual were placed under heavy scrutiny and followed. Those who were caught engaging in homosexual conduct or who confessed to it were put under pressure to name other women. Oral histories confirm that such investigations were a continuous part of military life throughout the entire period the women's services existed.

The 1964–65 witch-hunts noted by Ford could have been prompted by broader security investigations. In 1963 ASIO representatives noted that they had received a briefing in their Victorian regional office about the prevalence of homosexuality in the WRAAF and the WRAAC. The WRANS had also advised that 'instances of homosexuality among females have become evident'. The report, focused mostly on the WRAAF and the WRAAC, noted: 'Authorities expressing concern at the evidence of homosexuality among female members of both Services which involves females having access to classified information or who have in part of their Services been trained in occupations on their return to civilian employment, e.g. employment in cipher, signals or teleprinter duties within government departments.'[71]

While noting that homosexual acts between women were not illegal, the ASIO report called female homosexuality 'a state of perversion displaying a characteristic moral weakness rendering the individual concerned susceptible to other character weaknesses and from a security point of view extreme potential for compromise, exposure or pressure'. ASIO maintained that 'it does appear necessary that a restricted index of such persons must be maintained and that the Service Authorities should advise ASIO whenever such a deviate is discharged on these grounds'.[72] Historian Robert French notes that the issue of homosexuality and supposed links to espionage was discussed at the highest levels. In 1964, an Australian Government Cabinet Paper on 'Persons with Serious Character Defects as Security Risks' directed that 'a homosexual should not be employed … in a position where he may have or obtain access to highly classified information'.[73] French states that this directive certainly resulted in increased scrutiny of homosexuality in the armed forces.

In 1967 Dawn Jackson, who was then Director of the WRAAC, publicly acknowledged that homosexuality was a 'problem' occurring within the female services. It is clear that the WRAAC viewed lesbianism through the medical model at this time, as a condition that could potentially be treated. Jackson told the *Australian* newspaper that 'with the number of women we have, lesbianism must exist. Our officers are trained to watch for it, and we have methods of dealing with it. Doctors and padres play an important role here.'[74] Jackson did not elaborate further on the 'methods' that were used to deal with lesbianism. However, her reference to medical professionals and clergymen reveals that she was viewing homosexuality through a medical lens or as something that counselling from a religious figure could address. The archival trail is silent on any formal policy on dealing with lesbian and bisexual women at the time of Jackson's statement. Still, the available records and oral histories suggest the standard approach was to investigate, interrogate and discharge suspected lesbian and bisexual women (particularly those whose sexuality was considered fixed).

Investigations continued and accelerated in the 1970s and 1980s, even as mainstream society was becoming increasingly aware of homosexuality and there were campaigns in every state and territory for law reform and decriminalisation of homosexuality.[75] In one 1974 case, an allegation of lesbianism at a WRAAC barracks in the 3rd Military District saw an investigation take place the following year. Nine women were recommended for administrative discharge as a result. One woman, who stated she had a girlfriend in Canberra, gave a statement to the investigating officer that said: 'I believe a person's private life is their own affair [and] that the government should stay out of a person's bedroom. This could blow the top right off the camp scene in the Army. I know a lot of girls in the Army who are camp and if they discharge them all they won't have much of an Army left.' Captain B.G. Steele, the investigating officer, asserted that while four of these women 'have been only experimenting with lesbianism they could still influence other younger girls coming into the Army who, like themselves, could be interested in experimental lesbianism'.[76]

Such investigations ran across the three services and were not limited to the WRAAC or to any one state. Leading Aircraftwoman Delia Quigley remembers that 'there were a lot of witch-hunts in Canberra in the late seventies, early eighties to find the gays and kick them out'. Quigley comments: 'It became quite apparent that there were these little hidden groups of people that were hiding who they were because of their sexuality. And you know I started to hear stories about people being kicked out because of their sexuality and that there were actually witch-hunts to find people who were gay or lesbian.'[77]

The suggestion that homosexuality made women in the service vulnerable to blackmail was deeply offensive to many of the women who served at this time. Patriotism was a value emphasised during military service, and it had motivated many women to enlist. Julie Hendy, who was discharged from the WRAAC in 1968 on the grounds that her homosexuality might make her a security risk and subject to blackmail, points out how ridiculous this concept was:

> The whole point about being a lesbian in the Army was, they used to say, 'Well, you're subject to blackmail.' You're only subject to blackmail 'cause they make it illegal. But I had already confronted that idea. And I knew that if anyone ever did try to blackmail me, in the remote chance someone tried to blackmail, there's no way I would betray my country. You'd have to have rocks in your head really. And I would just front up and say, 'Look, this is happening and it's because of—' and I would've been discharged. But it's much better than the horror of going down in the other path. I'm just not a moral coward.[78]

Delia Quigley, reflecting on the way lesbians were treated as a potential threat to military security, points out the homophobia at the centre of this perception:

> There was all this stuff about you were a security risk, you know, because people could bribe you because of your sexuality, but when you think about it there was so many people getting drunk and, you know, could have been susceptible to bribery as well that they were probably just as much [as], if not more of a security risk.[79]

Viv Daniels, who joined the WRAAC in 1981, also identified this double standard. She was aware of married colleagues who were having affairs: 'Why wouldn't they be blackmailed too? Why is it any worse [to be gay]?'[80]

Lyndal Green, who joined the WRAAC in 1979, had direct experience of negotiating the witch-hunts that were directed towards women suspected of not being heterosexual. She had come to realise her sexuality relatively early, around the age of fourteen or fifteen. After completing her WRAAC training, she was posted to Puckapunyal and then to Sydney. She remembers Sydney in the early 1980s as being 'every lesbian's dream. I mean, god, pick a gay bar, any gay bar. It was like, go down Liverpool Street in Sydney, and it was unbelievable and no one gave a shit.' Although there were plenty of options for socialising around Sydney's 'gay golden mile', she was keenly aware of the consequences of having her sexuality exposed and the witch-hunts that were conducted: 'We'd seen and heard of girls that had just gone through shit, absolute shit. Have had their families annoyed, have been followed, have been photographed, friends have been interviewed, and so we just thought, "No, I don't want to go through that. This is just too bloody hard", and [we] need to keep it secret.'[81]

Ronni Spencer remembers that while she was a WRAN based at Watsons Bay, Sydney, she and her friends would frequent mixed bars because they were less likely to attract the suspicion of military police.[82] Socialising with other bisexual and gay men and women had to be done surreptitiously. Green remembers:

> This is before mobile phones and stuff. We had to work out signs and passwords to let people know when the military police were around, because they would sit outside the gay bars with cameras and take photos of you going in and coming out. So we would avoid certain areas, and it was all word of mouth and it was all done very quickly as much as you could, because it wasn't worth the angst. Just wasn't worth the angst.

The military police often followed Green, including on one occasion when she was travelling from Puckapunyal to Melbourne to visit her mother. She remembers pulling into her mother's driveway, 'and there's this bloody car behind me and I'm like "What?"' She was then questioned about who lived in the house she was visiting and how frequently she travelled to Melbourne. Green remembers, 'We didn't go to any gay bars in Melbourne when we were in Puckapunyal. It was all too close to home.'[83]

In 1977 Susie Struth was confronted by two Army policemen who showed up outside her house at 8.00 am. She remembers being asked: '"Are you gay? Are you a lesbian? Are you in a relationship?" The details of that day are very hazy except that we were there all day, and I can't remember why. It was quite intimidating. If intimidation was the factor, then it worked.' Her ordeal continued as she was made to attend a series of follow-up interviews at Victoria Barracks in Melbourne, and an official from the Special Investigation Branch kept trying to pressure her to name other women in her unit who were also gay. She remembers:

> [I] went into this room and there was me and a Lieutenant whose name escapes me, but it was—you imagine a single light bulb on a desk; that was it. He said, 'So, Corporal Struth, we want you to name every other lesbian that you know.' I just looked at him and I said, 'Oh, I don't think I can do that, Sir.' He said, 'No, we want you to.' I said, 'No, I think you taught me better than that.' I said, 'I'm not going to give that away. Sorry.'[84]

In 1979, Anna van Netten, who served with the WRAAC, was called before the military police when they learned she had been involved with another woman. She remembers receiving a warning, a 'don't do it again, don't get caught sort of thing'. She remembers it as a stressful and personally invasive experience: 'It was a harrowing afternoon having to face all these three blokes and a woman asking these really personal questions that I thought have got nothing to do with you guys. It just shouldn't happen.'[85] Like many other bisexual and lesbian women, van Netten had previously been deployed to a different location after military officials suspected she was engaging in homosexual conduct. This official warning was just a further tactic to let her know she was being monitored.

Those women who were caught up in the witch-hunts and interrogated or had their sexuality revealed usually faced discharge after undergoing interrogation. Occasionally, though, the women's services would attempt first to 'manage' lesbian behaviour, usually through deploying women who were in relationships to separate locations or by making women undergo psychiatric treatment. Ford has argued that 'relocation was one strategy for attempting to end lesbian relationships' in the military in the 1950s and 1960s.[86] Nola Strawbridge remembers seeing this happen during her time in the WRAAC in the 1960s. She describes one couple: 'They'd spent a lot of money training them, but they'd post them off somewhere and hope the relationship would stop there.'[87] Jennifer Jefferies was investigated and was separated from her partner in the 1970s. She recalls: 'They didn't kick us out, but they posted us to the opposite ends of Australia.' She thinks this was a deliberate strategy: 'Without a doubt. They split us up. They very much split us up.'[88]

In 1973 an anonymous WRAAF who was discharged on the grounds of her sexuality told *CAMP Ink* that she believed that if she had requested to 'see a psychiatrist because I was falling in love with a girl, then the authorities wouldn't have touched me'.[89] It was her resolution that her sexuality was immutable that conflicted with military policy. Paperwork from a 1979 RAAF investigation shows that women found to have engaged in homosexual conduct were asked various questions in their interrogations to ascertain the 'degree' of their homosexuality. In this case, the woman being investigated was asked numerous detailed questions about sexual practices and whether she had played the 'active' or 'passive' role. She was also asked, 'Do you consider yourself completely lesbian or bisexual?' When she affirmed that she was 'completely lesbian', she was subsequently asked, 'Do you want to stay this way?' She replied, 'Yes.' The paperwork from this investigation notes that the woman had been sexually assaulted as a teenager. A fellow servicewoman who was asked for supporting information about the woman under investigation told military officers that 'from discussion with me it appears to me that if she was given medical assistance she may be cured of the problem'.[90] This reveals that even after the American Psychiatric Association removed homosexuality as a disorder from its globally used *Diagnostic and Statistical Manual of Mental Disorders* in 1973, the Australian military still viewed female (and male) homosexuality through this prism.

Of course, not everyone in the armed forces supported the policy on homosexuality, and some were even able to support or protect their friends. Gen Ford remembers that she came out to about five or six of her closest Army confidants, and 'Without exception really they were all very supportive and said, "Doesn't bother me, Gen. It's absolutely, it's cool. No problem."'[91] Chris Sinclair even recalls one male commanding officer who was very conscious that there were lesbian soldiers serving under him, but he actually preferred them. She explains:

> I'm positive that the captain knew that I was gay; however, [he made] comments such as 'I don't care if they're gay', 'I don't care; I get better work out of one female compared to ten guys. Women are more mature—they have a relationship break-up they don't fall apart.' And the brigadier once said to me, he said, 'I don't care.' He said, 'I'd rather those [lesbians] because they're here for a career and not to get pregnant and go off and nah, nah, nah, nah.'[92]

Of course, servicewomen never knew who would be friend or foe and had to be very cautious about to whom they disclosed their sexuality. Moreover, once an investigation was triggered, there were limits to how much allies could protect servicewomen from the consequences.

Discharges and negotiating trauma

In 1966 Sandra, who had served with the WRANS for a year, was discharged with the dishonourable grounds of 'services no longer required'. Her discharge was issued after the discovery of a relationship she had been in with a fellow servicewoman. The two women had been seeing each other for around three months before Sandra was warned by a fellow servicewoman to 'be careful. They're watching you.' After both women applied for a three-day leave pass for the same days, they were confined to barracks without explanation before being interrogated separately about their relationship. After writing down a statement, Sandra was 'shipped' out of Darwin two days later. Once she had arrived in Melbourne, she was sent to HMAS Lonsdale, where she was informed of the decision. She was told that she was being discharged 'because they considered me a security risk because I could be blackmailed into giving some secret away'.[93]

Carole Popham and Christina Dennis, who met and formed a relationship in 1967 while serving in the WRAAF, chose to disclose their relationship and leave voluntarily. This decision had been exacerbated by the women being posted to separate locations and Popham's premonition that witch-hunts were about to start again. This turned out to be an accurate premonition, as she believes that around a dozen women were forced out of the WRAAF in 1968 on the basis of their sexuality. One of the benefits of requesting their own discharge was that the discharge certificate simply stated 'On request', rather than 'services no longer required', which was the certificate usually given to women in the WRAACS and WRANS (such as Sandra) who were discharged on the grounds of homosexual conduct. Popham explains, 'The worst discharge, the dishonourable discharge, is "services no longer required". The one that's nice is "On Request". They just don't bother to say on whose request.'[94]

Julie Hendy had clearly been considered one of the best and brightest in the WRAAC before her sexuality was discovered. In December 1967, as an Australian corporal, she was part of a small and select group of ten women sent to Singapore. They were the first members of the WRAAC to serve in an overseas theatre since the formation of the corps. In 1968 when she was caught in an intimate position with a British servicewoman in Singapore, she was quickly sent back to Australia and discharged with a certificate stamped 'Retention in the military forces not being in the interests of those Forces'. Hendy contrasts her experience with a woman she knew who had served and had been caught in bed with a man, noting that this woman had been discharged at her own request. This was a marked difference from the certificate Hendy was issued.[95]

The threat of being issued a 'services no longer required' discharge caused considerable angst to many in the military. Delia Quigley was aware that 'there was some quite nasty stuff going on with dobbing other people in so that people wouldn't get kicked out themselves and because people didn't want to have a dishonourable discharge'. Being discharged with 'that black mark against their name' would often dramatically impede the ability of ex-servicewomen and men to find employment after their time in the military. Quigley eventually found herself called before her warrant officer after a civilian who worked in the same communications centre saw her at a party at the weekend and told her boss 'that she thought I was a lesbian'. Quigley remembers the warrant officer saying, '"I believe that you might be a lesbian. And what are we going to do about it?" And I just said to him, "Well, I've only got a couple of months left on my sign-up period. Can I just keep my head down and quietly go when my sign-up period is done?"'[96] The warrant officer agreed to this, and Quigley left as soon as her sign-up period finished.

Susie Struth, who had been severely interrogated within the WRAAC, asked her commanding officer if it was possible to fight the investigation into her sexuality but was told it was not. She was given the option of being dishonourably discharged, being discharged on the grounds of not being suitable for the military or serving out the remaining time of her service period and leaving. These options made it clear that there was no future for her in the Army. Today, Struth wishes she had contested this more but believes that leaving was ultimately the right decision for her as she did not wish to have to hide her sexuality.[97]

Being discharged from the services was a traumatic experience for many. Even the most resilient of women were affected by the process. Julie Hendy, who later went on to enjoy a successful and distinguished career outside the military, remembers the message conveyed to her after she was identified as a lesbian was that 'I'm in total disgrace, yes, and I've betrayed the country. And I have to be sent home in disgrace.' She struggled with being someone who had risen to such a prized position within the Army to being dismissed so rapidly: 'When you've been top of what you're doing and they send you over to Singapore and then they say, "Oh clearly no, you're rubbish" and get rid of you.' She emphasises that an experience like that would certainly make someone feel 'pretty clear that you're not quite right'. She continues: 'And it was a mental illness and blah, blah, blah until 1973. And I believed it. And it wasn't for a couple of years that I began to slowly piece together the fact that no, I'm actually quite a nice person. And—but it's still there. There's residual scars, I think, if I look at it.'[98]

Sandra is still affected today by what she experienced. She states: 'I don't cry. I don't know why. I don't cry, do I? I don't know whether that—all of that—has affected my emotions. I just keep it inside, just keep it all inside … so you get emotionally stunted. And that's—yeah, it's just because of the way you're treated back then, and then later.'[99]

Conclusion

From the 1950s to 1985, the women's services provided an environment in which an extensive, yet clandestine lesbian subculture thrived. Many women discovered and explored their sexuality in this environment. From the very start, though, the services were aware that lesbian and bisexual women were attracted to the military and implemented practices that became colloquially known as witch-hunts, intended to force non-heterosexual women out of the military. Intense interrogations often left women with long-lasting psychological trauma. As chapter 3 will show, the treatment that was meted out to lesbian and bisexual women from the 1950s onwards formed the prototype for policies and procedures widened to persecute gay and bisexual men as well after 1974.

By forcing bisexual and lesbian women out of the women's services, the Australian military lost a great many talented and dedicated servicewomen. A significant number of bisexual and lesbian servicewomen who managed to conceal their sexuality also opted to leave prematurely, believing that they would either be found out eventually or that their sexuality would hold them back. Jennifer Jefferies feels she was denied opportunities in the WRAAC because superior officers were aware of her sexuality: 'So I was never going to get a decent posting. I was never going to get a decent promotion. Just—it killed my career.' Still, leaving a career she loved was very difficult: 'It broke my heart to leave the Army. It really did. That had been a big dream.'[100] Anna van Netten, who had joined the WRAAC aged seventeen in 1973, left voluntarily after ten years of service at the rank of lance corporal. She believed that official suspicion about her sexuality prevented her from advancing. Reflecting on the military, van Netten is positive about what the institution offered her: 'I loved the Army. I loved the whole time that I was there, right up until the day I got out. I still love it. It's a great career and I recommend it to anyone, whether they be straight or gay—it doesn't matter.'[101]

Notes

1 Janet Hawley, 'Closet lives', *Age Good Weekend*, 9 November 1996, p. 25.
2 Ford, 'Disciplined, punished and resisting bodies', pp. 53–70.
3 Robinson, 'Sexuality at a cost', pp. 45–58.
4 Riseman, Robinson and Willett, *Serving in Silence?*, p. 64.
5 Kate Darian-Smith, 'War and Australian society', in *Australia's War, 1939–45*, ed. Joan Beaumont (Sydney: Allen & Unwin, 1996), p. 62.
6 See Janette Bomford, *Soldiers of the Queen: Women in the Australian Army* (South Melbourne: Oxford University Press, 2001) and Margaret Curtis-Otter, *WRANS: The Women's Royal Australian Naval Service* (Sydney: Naval Historical Society of Australia, 1996).
7 Ashley Ekins, 'Vietnam: A winnable war?' in *War Strategy and History: Essays in Honour of Professor Robert O'Neill*, ed. Daniel Marston and Tamara Leahy (Canberra: ANU Press, 2016), pp. 5–30.
8 Ford, 'Lesbians and loose women', pp. 81–2.
9 'NSW woman in army post', *Daily Telegraph* (Sydney), 11 November 1941, p. 5.
10 Ford, 'Disciplined, punished and resisting bodies', p. 56.
11 Ibid.
12 Bomford, *Soldiers of the Queen*, p. 1.
13 Lapovsky Kennedy and Davis, *Boots of Leather, Slippers of Gold*, pp. 77–8.
14 'Women puzzle the army', *News* (Adelaide), 13 November 1953, p. 2.
15 See for example *News* (Adelaide), 13 November 1953, p. 2; *Brisbane Telegraph*, 19 November 1953, p. 23; *West Australian*, 14 November 1953, p. 3.
16 Emily Wilson, '"Someone who is sick and in need of help": Medical attitudes towards homosexuality in Australia, 1960–79', in *Homophobia: An Australian History*, ed. Shirleene Robinson (Sydney: Federation Press, 2008), pp. 148–71.

17 Jennings, *Unnamed Desires*; Reynolds and Robinson, *Gay and Lesbian, Then and Now*, pp. 7–10.
18 Robert Reynolds, *From Camp to Queer: Re-making the Australian Homosexual* (Melbourne: Melbourne University Press, 2002), pp. 1–45.
19 Ibid., p. 60.
20 Ford, 'Disciplined, punished and resisting bodies', p. 60.
21 'Transcript of segment of Mike Willesee's "A Current Affair"', CTC 7, 17 July 1973, available from the WRAAF's service record. The archival reference is not provided here because it identifies the individual concerned. The person might still be alive and might not want their sexuality disclosed. Please contact the authors for further information if required.
22 WRAAF's service record.
23 Reynolds and Robinson, *Gay and Lesbian, Then and Now*, pp. 63–5.
24 'Policy on homosexuality in the services', October 1976, NAA A7481, 22.
25 Robinson, 'Sexuality at a cost', p. 49.
26 'Department "will not accept" homosexual behaviour', *Canberra Times*, 21 January 1982, p. 11.
27 'Women essential part of the Australian Army', *Mercury* (Hobart), 18 January 1952, p. 5.
28 'They wear the air force blue', *Advertiser* (Adelaide), 14 June 1953 p. 11.
29 'WRANS doing fine job', *Argus* (Melbourne), 14 October 1955, p. 22.
30 Bomford, *Soldiers of the Queen*, p. 4.
31 'Smart uniforms and nylons for girls in navy', *News* (Adelaide), 2 April 1951, p. 13.
32 Kathryn Spurling, 'From exclusion to submarines: The integration of Australian naval women volunteers', *Australian Defence Force Journal*, no. 139 (November–December 1999): 37.
33 *Argus* (Melbourne), 6 March 1952, p. 14.
34 'Girls who wear the air force blue', *Advertiser* (Adelaide), 22 June 1954, p. 7.
35 Bomford, *Soldiers of the Queen*, p. 75.
36 Spurling, 'From exclusion to submarines', p. 37.
37 Noah Riseman and Richard Trembath, *Defending Country: Aboriginal and Torres Strait Islander Military Service since 1945* (St Lucia, Qld: University of Queensland Press, 2016).
38 Riseman, Robinson and Willett, *Serving in Silence?*, p. 68.
39 Bomford, *Soldiers of the Queen*, p. 117.
40 Nola Strawbridge, interview with Shirleene Robinson, 15 May 2016, Sydney.
41 Carole Popham and Christina Dennis, interview with Shirleene Robinson, 18 March 2016, Melbourne.
42 Julie Hendy, interview with Shirleene Robinson, 13 May 2016, New South Wales.
43 *Seven Days: Love is Love*, 15 February 1966, Channel Seven; *Sydney Morning Herald*, 20 February 1966, p. 42.
44 Elaine Crump, interview with Noah Riseman, 23 August 2017, Brisbane.
45 Lyndal Green, interview with Shirleene Robinson, 1 July 2016, Ballarat.
46 Ronni Spencer, interview with Noah Riseman, 3 August 2017, Melbourne.
47 Raymond Evans and Kay Saunders, 'The evolution of the Australian housewife', in *Gender Relations in Australia: Domination and Negotiation*, eds. Kay Saunders and Raymond Evans (Sydney: Harcourt, Brace, 1992), p. 191.
48 Julie Ustinoff, 'Homebodies and weekend handymen', in *The 1960s in Australia: People, Power and Politics*, eds. Shirleene Robinson and Julie Ustinoff (Newcastle upon Tyne: Cambridge Scholars Publishing, 2012), p. 148.
49 Michelle Arrow, *The Seventies: The Personal, the Political and the Making of Modern Australia* (Sydney: NewSouth, 2019), p. 232.

50 Hawley, 'Closet lives', p. 25.
51 'Pencil', interview with Shirleene Robinson, 9 March 2017, Castlemaine, Vic.
52 Nola Strawbridge, interview with Shirleene Robinson, 15 October 2015, Bribie Island, Qld.
53 Nola Strawbridge, interview, 15 May 2016.
54 Sandra, interview with Shirleene Robinson, 3 April 2017, Orbost, Vic.
55 Susie Struth, interview with Shirleene Robinson, 18 February 2017, Melbourne. See also Riseman, Robinson and Willett, *Serving in Silence?*, p. 103.
56 Delia Quigley, interview with Shirleene Robinson, 21 May 2017, Canberra.
57 Jennings, *Unnamed Desires*, p. xi.
58 Rebecca Jennings, '"It was quite a scary time": Lesbians and violence in post-war Australia', in *Gender Violence in Australia: Historical Perspectives*, ed. Alana Piper and Ana Stevenson (Clayton, Vic: Monash University Publishing, 2019), p. 147.
59 Julie Hendy, interview. See also Riseman, Robinson and Willett, *Serving in Silence?*, pp. 63–6.
60 Ford, 'Discipline, punished and resisting bodies', p. 55.
61 Jennifer Jefferies, interview with Shirleene Robinson, 12 September 2016, Sydney.
62 Lyndal Green, interview.
63 Delia Quigley, interview.
64 Chris Sinclair, interview with Noah Riseman, 10 September 2018, Melbourne.
65 Lyndal Green, interview.
66 Gen Ford, interview with Noah Riseman, 13 November 2014, London.
67 Elaine Crump, interview.
68 Sandy Ashton, interview with Noah Riseman, 29 August 2015, Melbourne.
69 Jenny Robertson, interview with Shirleene Robinson, 24 May 2017, Hobart.
70 See also Riseman, 'Hunting gays and lesbians in the Australian Defence Force, 1974–1992', pp. 325–56.
71 'Security Personnel Personal Characteristics Including Homosexuality Consideration as Security Factor', vol. 1, November 1963, NAA A6122, 1194.
72 Ibid.
73 Robert French, '"Persons with serious character defects": Homosexuals in the Commonwealth Public Service, 1953–1974', in *Intimacy, Violence and Activism: Gay and Lesbian Perspectives on Australasian History and Society*, ed. Graham Willett and Yorick Smaal (Clayton, Vic: Monash University Publishing, 2013), p. 127.
74 *Australian*, 27 September 1967, p. 9.
75 Reynolds and Robinson, *Gay and Lesbian, Then and Now*, pp. 63–6; Riseman, 'Hunting gays and lesbians in the Australian Defence Force, 1974–1992', pp. 325–56.
76 'Investigation into Alleged Homosexual Activities', Department of Defence (Army Office), 7 January 1975, NAA A7481, 22.
77 Delia Quigley, interview.
78 Julie Hendy, interview. See also Riseman, Robinson and Willett, *Serving in Silence?*, p. 71.
79 Delia Quigley, interview.
80 Viv Daniels, interview with Shirleene Robinson, 5 August 2016, Melbourne.
81 Lyndal Green, interview.
82 Ronni Spencer, interview.
83 Lyndal Green, interview.
84 Susie Struth, interview. See also Riseman, Robinson and Willett, *Serving in Silence?*, pp. 112–13.
85 Anna van Netten, interview with Shirleene Robinson, 6 April 2016, Newcastle.
86 Ford, 'Disciplined, punished and resisting', p. 57.

87 Nola Strawbridge, interview, 15 May 2016.
88 Jennifer Jefferies, interview.
89 'So you want to be a WRAAF', *CAMP Ink* 3, no. 2 (1973): 5.
90 NAA ST6999/1, 705/11/P1 Part 1.
91 Gen Ford, interview.
92 Chris Sinclair, interview.
93 Sandra, interview.
94 Carole Popham and Christina Dennis, interview. See also Riseman, Robinson and Willett, *Serving in Silence?*, pp. 50–3.
95 Julie Hendy, interview. See also Riseman, Robinson and Willett, *Serving in Silence?*, pp. 69–72.
96 Delia Quigley, interview.
97 Susie Struth, interview. See also Riseman, Robinson and Willett, *Serving in Silence?*, p. 114.
98 Julie Hendy, interview. See also Riseman, Robinson and Willett, *Serving in Silence?*, p. 73.
99 Sandra, interview.
100 Jennifer Jefferies, interview.
101 Anna van Netten, interview.

Chapter 3

The heightened ban years, 1974–92

In 1973, the gay and lesbian newsletter *CAMP Ink* published an interview with a recently dismissed lesbian airwoman entitled 'So you want to be a WRAAF'. The article outlined how a friend had warned the anonymous WRAAF at the centre of the story that RAAF Police were contacting a few servicewomen. The next day the RAAF Police called the servicewoman into an interview. Over the course of eight hours the police asked about her relationships, the names of other lesbian WRAAFs, and questions such as 'if I looked at a woman do I have an orgasm [*sic*], that type of thing. And I said, "You men might but I don't."' The police wrote a statement on behalf of the WRAAF and, exhausted, she signed it. Then they accompanied her to her residence and searched her room, confiscating some letters from her girlfriend. Two weeks later the WRAAF was given the opportunity to request her own discharge. She agreed, fearing that otherwise 'They probably would have discharged me by stating "services no longer required", which they did to one pal of mine I met the other night'.[1]

The article in *CAMP Ink* had significant ripple effects. The mainstream media picked up the story when a second dismissed WRAAF wrote to the Defence Minister, Lance Barnard.[2] This was the era of the Whitlam Labor government, which was more sympathetic to homosexuality than previous Australian governments. Barnard ordered the three services to review their procedures for investigating suspected homosexuals and to adopt a consistent procedure that 'should be broad, understanding and designed to cause the least embarrassment in such situations whilst safeguarding the interests of the Service'.[3] In 1974, this review recommended the first consistent, tri-service policy to ban and investigate suspected lesbians, gays and bisexuals. The services occasionally updated the policy—most notably in 1985—but the general set of procedures for investigating and discharging LGB people remained until 23 November 1992.

With a clear set of guidelines in place, and with homosexuality a more visible topic of public debate, a process was set to target LGB people within the ADF. This chapter juxtaposes ADF policy documents and investigation reports alongside oral histories of LGB people who were investigated and discharged during the period from 1974 to 1992. There is a strong, consistent set of tropes that LGB ex-service members recollect: interrogations that could continue for hours on end; graphic and invasive questions about sex lives and relationships; pressure to name other LGB service members; and the compulsion to request their own discharge rather than receive a dishonourable discharge. Investigations followed suspects' social networks to uncover other LGB service members. Colloquially, these investigations became known as witch-hunts.

Notwithstanding the risks and investigations, LGB service members still found ways to explore their sexuality. For male officers and women of all ranks, there were LGB social networks and subcultures across the three services. Men from the other ranks sometimes found other gay or bisexual servicemen for sex, but most oral histories suggest that they tended to turn to civilian life for sexual opportunities and relationships. By the late 1970s, in Australian cities there was a growing number of LGB bars, sex clubs, social groups and other organisations. LGB members took advantage of these opportunities while ever-wary of the watchful eye of service police, who were even known to infiltrate LGB venues undercover. It seemed that as society was slowly growing more tolerant of homosexuality, the ADF was in turn becoming more heavy-handed in enforcing its ban.

The policy framework

The 1974 review of the services' policies towards homosexuality acknowledged that there was pressure within civilian society to decriminalise male homosexual acts. Indeed, South Australia became the first state to do so in 1975, followed by the Australian Capital Territory in 1976, Victoria in 1980, all other mainland jurisdictions by 1990 and Tasmania in 1997.[4] Meanwhile, lesbianism, which had never been a criminal offence, was becoming more visible and gaining greater social acceptance. Yet the 1974 tri-service review argued that the services should not implement reforms relating to homosexuality because:

1 The presence of LGB members would disrupt command and disciplinary relationships.

2 The armed forces' reputation would decline, adversely affecting troop morale, recruitment and retention.
3 Even if states implemented law reform, there was still a social stigma attached to homosexuality that put LGB people at great risk of blackmail.
4 Homosexuality was a threat to minors in the armed forces, whom the services had a legal obligation to protect.

One justification bizarrely talked about the ban protecting LGB people from bullying: 'within barracks, any member with known homosexual tendencies would be made the butt of sarcasm and ill humour, and would be unable to gain the respect of fellow personnel'.[5] It was an early example of blaming the bullied, rather than the organisation considering programs and leadership to tackle discrimination and harassment.

The new policy introduced step-by-step procedures for investigating cases of homosexuality. First, any suspected homosexual should be reported to the commanding officer. If there were a *prima facie* case, the commanding officer should refer the matter to the service police to investigate. The minority of cases that involved sexual assault, a senior and junior rank, minors or indecent acts in public were to be investigated as disciplinary matters under military law. All other cases were to be dealt with administratively. Service police were responsible for gathering evidence and ultimately determining whether the homosexual act was an isolated incident or whether the person were a 'confirmed homosexual'. If investigators considered it to be an isolated homosexual act, the Defence member could be retained with a stern warning. 'Confirmed homosexuals' would have the choice either to request an honourable administrative discharge 'at own request' or else they would receive a dishonourable discharge 'services no longer required' or 'unsuited for further service'.[6]

The policy advised that throughout the investigation process, suspected homosexuals should be treated 'with sympathy and discretion', and ADF statements for the next eighteen years regularly used this phrase to describe the organisation's treatment of homosexuals. The investigation procedures placed restrictions on police methods, which were designed to protect suspects' rights. For instance, police were expressly forbidden from searching property without the suspect's consent. The policy stipulated that female police should always be present in interviews with servicewomen. All suspects had the right to request that an officer be present during police interrogations. Importantly, the rules imposed limits on the scope of interviews:

> Questions may be directed to establish the circumstances of the case, identify others involved and ascertain whether action on

> related matters, such as possible compromise on security, is required. Questions on the detail of sexual acts is to be avoided except to the minimum necessary to establish that homosexual conduct has in fact occurred and that the person concerned fully understands the nature of the allegations.[7]

As the following examples reveal, police almost always considered it necessary to ask about sexual acts to prove that homosexual conduct had occurred. Moreover, interview participants regularly recall significant police pressure to name other LGB Defence members.

The most significant update to the Defence LGB ban was in November 1985 when the Chief of the Defence Force (CDF) approved Defence Instruction (General) PERS 15-3: 'Homosexual Behaviour in the Australian Defence Force'. The updated order affirmed that administrative discharges for homosexuality did not breach the *Defence Force Discipline Act 1982* (Cwlth). The investigation procedures were relatively unchanged from the 1974 policy. The only major difference was an added justification for the LGB ban: 'Health: The health risks to persons engaged in homosexual activity are real and well documented.'[8] This was an allusion to HIV and AIDS, and drew on stereotypes of gay men as carriers of the virus. As chapter 4 outlines, this and the other justifications for the ban came under increasing attack in the late 1980s and early 1990s. What also came under attack was the way the ban operated in practice, with heavy-handed investigation tactics that rarely constituted 'sympathy and discretion'.

Starting investigations

All sorts of behaviour could arouse suspicion that a service member was lesbian, gay or bisexual. Ways of walking, talking, mannerisms or hobbies generally played on contemporaneous constructs of masculinity or femininity. Those men who exhibited stereotypes associated with femininity and women who manifested masculine attributes were easy targets of colleagues, commanding officers and service police. Being associated with someone who was suspected or proven to be LGB also could result in someone being subjected to investigation. Sometimes, someone spotted service members off base either at a known LGB venue or with a partner. A tip-off to the service police would then spark an investigation. Tip-offs could come from different sources and for various reasons. 'John' recalls serving in north Queensland in the mid-1970s when a soldier's gay friend kept ringing the base and was distraught when the soldier was unavailable. That soldier was soon outed and discharged.[9]

In 1991 an anonymous ex-RAAF policeman wrote a long letter to the gay magazine *OutRage* detailing service police practices. He also explained: 'Any Service member who doesn't like another member for any reason can report that person to the PROVOST UNIT, the SPs ("Spits") as having made a sexual advance to him or her. Even an anonymous note or telephone message can find the recipient of such an accusation DISMISSED FROM THE SERVICE if the follow-up Service Police investigation bears "evidence".'[10] To give one example: in 1980 WRAAC member 'Pencil' was falsely accused of making advances towards a junior recruit who had attempted suicide. Under threat of an investigation and dishonourable discharge, a devastated Pencil nearly resigned until some supportive friends persuaded her to fight the charge. She was ultimately successful and wound up serving a total of twenty-five years.[11] It was not until 1992 that the ADF even proposed mechanisms to address accusations that were 'false and malicious, vexatious or mischievous'.[12]

Women were especially vulnerable to be reported to police if they were seen to be challenging male authority or harassment, which at times could be endemic.[13] In 1987 'Terri' was one of only two airwomen who worked as Telus technicians. She was based at RAAF Base Glenbrook in western Sydney, where she was subjected to bullying and sexual harassment from her superiors. She even recalls attempted sexual assaults: 'He pinned me up against walls; he threatened to strangle me. He's the type of person that would say over and over again, "Oh, you only just need a good root"—I think would have been the word he used back then. You know, "That's all you need and I'm the person to do that for you."' She broke a superior's arm when refusing one advance, and he subsequently reported Terri for being a lesbian.[14]

Drug investigations were another opening that could unexpectedly uncover LGB members. There was not a link between drugs and homosexuality *per se*; rather, there are documented cases in which suspects in drug cases named gays or lesbians in an effort to deflect the investigation. For instance, in March 1989 a suspect in a routine drug interview with the Army Special Investigation Branch (SIB) made allegations of homosexuality against at least nine other soldiers. The subsequent investigation uncovered at least one bisexual, two gay men and a soldier in a relationship with a transgender woman.[15] In 1988 both Yvonne Sillett and 'Mark' were serving in Signal Corps at Watsonia Barracks, Melbourne. A suspect in a drugs investigation named Sillett as a lesbian. She refused to comment during her police interview, and her security clearance was downgraded. No longer allowed to perform her role, Sillett was compelled to request her own discharge.[16] SIB called Mark into an interview because they suspected him of drug use. As the intimidating questions continued, Mark eventually cracked and announced that he was gay.[17]

Sometimes LGB people were discovered completely by accident. Janet Carter had undergone six months of a gruelling selection process before she was accepted to undertake Army officer training at Duntroon in 1986. In her final days of training, she remembers being in a class when she was mysteriously summoned to meet two strangers at the mess tent. They identified themselves as being from SIB. Carter says: 'And I remember at that moment just looking at his face while I'm shaking his hand, I have no idea what I've done but whatever I've done I'm fucked. You know. Yeah. Because from the start it was clear—you know, military police were bad enough. Special Investigations Branch [were] the equivalent, the Army equivalent, of Criminal Investigations.' Carter later found out that her ex-girlfriend, who was also at Duntroon, had attracted the attention of superiors after she had missed morning parade and had been reported AWL. Officers searched her room and found a letter from Carter along with a photograph of her. Carter asserts, 'So if there hadn't been a photograph I might have got away with it. The fact that she had the photograph and the letters together implicated me.' The consequences of this discovery were grim for Janet Carter, who remembers, 'I went from golden girl to pariah within the space of a few hours.'[18]

There were some instances when LGB people voluntarily came out so that they could discharge. One of three cases of homosexuality reported in a July 1982 summary of offences in the RAAF was a 'Member [who] made admissions of homosexuality when applying for a discharge at own request'.[19] In 1980 Senior WRAN Elaine Crump told her divisional officer that she was gay and wanted to leave the Navy. Officials revoked her security clearance and sent her to Cerberus, where she had a short interview with WRANS regulators—the WRANS equivalent of Naval Police. Crump's discharge at her own request came through a few months later.[20]

Leading Aircraftman Kyle Duffield was also tired of hiding his homosexuality and outed himself in 1989.[21] The 1974 policy stated that when someone came forward as homosexual, 'the investigation is ordinarily to be confined to ascertaining whether the admission is true. However special circumstances, such as the involvement of other service personnel, may make fuller investigation necessary.'[22] Duffield's police interview transcript reveals that police still asked the standard questions about his sex life and relations, and were very much trying to extract the names of other gay servicemen.

Of course, commanding officers or members of the service police could exercise discretion when deciding whether or how to initiate investigations for homosexuality. Service police who were themselves LGB or who were straight allies could sometimes forewarn someone about an investigation. In some circumstances they could protect their friends and prevent an investigation from proceeding. Kate Carlisle recalls that, when someone reported her as a lesbian to HMAS Harman around 1989–90, 'I got a phone call from a petty officer, who was a good friend of mine, who was a gay girl, saying, "Kate, just letting you know, this letter's coming." I said, "Oh, shit. What am I meant to do?" She said, "Don't worry about it. I've destroyed it."'[23] Of course, not every LGB member was so fortunate, and for them the investigations followed similar patterns that often culminated in their discharge.

Surveillance

Service police generally needed evidence before they could confront suspected LGB members. Interviewing the accused's social networks was one method, and the other main technique was to monitor suspects closely. Surveillance could uncover evidence of suspects visiting LGB venues, or observing them in compromising, affectionate situations with partners. Of course, police tried their best to be discreet. Tony McLeod learned only after he left the Army in 1989 that SIB had parked cars outside his house and that agents were following him.[24]

At other times the police presence was obvious to the LGB members. In 1981 Shane Duniam was living in Williamstown in Melbourne and stationed at RAAF Base Point Cook. He recalls: 'I used to go out quite a lot then, clubbing, having a damn fine time, and then I would notice leaving Point Cook headlights behind me, that would follow me. Or during the day a car would follow me. And then I realised that something was up.' Duniam was fed up with the surveillance and cracked one night: 'I drove off, and as I got closer I saw them sink down in the seat. I went "Fuck this!" and I just put my brakes on, wound down my window: "Hey, fellas, I'm going to Mandate [gay nightclub] tonight. I'm probably home about 3.00 in the morning. You can have the night off."' A few days later Duniam was called in for an interview and confessed to being gay.[25]

As the decade went on, service police—especially in the RAAF—became more assertive in their surveillance. In April 1991 *OutRage* published an article exposing RAAF Police tactics with information provided by an anonymous officer. The article caused a stir because it included the names, ranks and service numbers of three RAAF police members who worked undercover to expose LGB members. The article explained that police practice was to place suspects under twenty-four-hour surveillance for two weeks. RAAF Police photographed all contacts and followed the suspects. In other cases, undercover police would visit LGB establishments and try to befriend and entrap service members.[26] Danny Liversidge was one person caught in this sort of investigation in 1991. RAAF Police had photos of Liversidge outside gay nightclubs and his then lover's house. At the end of the pile was 'a photograph they put down and it was actually me kissing a guy. It was Darren, standing outside his [housing] unit. I was obviously heading back to the base and they had a photograph of me kissing a guy. It was like—and then the question came out: "Are you a homosexual?"' Liversidge confessed and requested his own discharge.[27]

The *OutRage* article also exposed search methods conducted by RAAF police. It described the investigation of 'Paul': 'A team of RAAF police ransacked his workplace, his car and his flat. "They went through everything: the freezer, my clothes, my dirty laundry. That was the worst two or three hours I have ever been through."' Police confiscated a few gay magazines and a novelty bong, and required a urine sample. Legal officers assisted Paul, and the search was ultimately ruled inadmissible because police did not have a warrant.[28]

Others who did not have legal advice were less fortunate, such as Shane Duniam. He described a similarly aggressive search in 1981 where police seized private letters written by another gay airman. Duniam remembers, 'When they finished searching my bedroom they went, "Where's all your pornography and stuff?" And I said, "What do you mean by stuff?" Knowing full well. And anyway, they said, "Oh toys", and I went, "Well, I kind of think that if it's not real it doesn't exist in my life."'[29] What Duniam did not know was that, according to a redacted record relating to his investigation, RAAF police believed that he and three other airmen were also involved in 'pornography sessions'. Upon reading this document, Duniam wrote: 'If it wasn't so sad it would be funny! … During my time in the Air Force I never witnessed a porn party (not that I would have gone anyway), or was invited or heard of such a thing.'[30]

The 1974 policy explicitly stated that service police could conduct searches only with suspects' consent. The above examples show how consent could often be compelled, and even ADF authorities recognised that this was problematic. In 1986, after a Defence member lodged a redress of grievance, a RAAF cable explicitly advised of updated rules: 'No power of search exists in respect of homosexual investigations, even with consent.'[31] Yet searches continued. The anonymous RAAF ex-policeman wrote to *OutRage* in 1991:

> Secret room checks and searches of personal belongings are regularly done with master keys and knowledge of the member's work schedules and social habits. The 'Spits' are authorised to enter any Commonwealth property, barracks or off-base billet, without a search warrant. The SPs are looking for homosexual pornography, letters from same-sex lovers, snapshots showing, say, two men in any kind of 'socially acceptable or otherwise' embrace—one's arm around another's shoulders at a function.[32]

Karl Bryant was a clerk in the RAAF and remembers reading documents about police searching mail and applying for approvals to search suspected homosexuals' houses. Bryant, too, came under suspicion when an employee spotted him at the movies with his partner in 1990. A friendly flight lieutenant warned Bryant that an investigation was imminent and advised him to go home and hide any evidence of his relationship.[33] Army Corporal Chris Sinclair remembers returning from Christmas and New Year's break in 1981 and thinking, 'Somebody's been in my room. I went my shoes, my boots were moved by an inch, and I went: "Somebody's been in my room." And then the next day I reported for work and I was told there was someone there to speak to me.' It was the RAAF Police. They were not after Sinclair but her friend Dennis, who was subsequently dismissed for being gay.[34]

These practices went unchallenged until the public exposure of the *OutRage* article caused a stir. Not only had it exposed practices that went beyond policy rules but also the revelation of the names and service numbers of RAAF Police constituted a security breach. The CDF, General Peter Gration, was in an awkward position because he knew the identity of the article's main informant: his son Richard, a RAAF pilot. The CDF ordered that service police rein in their practices; rather than going out searching for LGB members, they were only to investigate cases brought to their attention.[35]

Interviews

By far the most intimidating aspect of service police investigations was the police interview. These could go on for hours or even days, and usually only ended when the suspect confessed to being homosexual. Keith Drew angrily recalls his SIB interview in 1980: 'I was told nothing, and then they started pulling me out of my battalion headquarters fucking once a day for a week, two weeks ... and they kept on asking me if I was gay, and I went, "Yes, I am", and that was it.'[36] Eunice Coogan, caught in 1982, vividly remembers her RAAF Police interview:

> For twenty-four hours I didn't tell them [I was a lesbian]. I wasn't going to say it. And some of the things that they've started to say to me were just terrible, absolutely terrible. And then they started the tactic of, 'Well, let's kind of like harass her in to saying things'. So you know, I'm busting to go to the toilet and then I've got to have one of them standing in front of me, watching me go to the toilet. Or I'm going to eat and they march me down—two policewomen standing beside me while I eat and they don't eat. So they are portraying themselves to be guards. Do you know? Instead of

> at least even having something to eat. So I have to go through all of this humiliation. And then, and then eventually I told them. I just couldn't take any more. Like it's this—you know, they just seemed hell bent on getting me to say it and that's it.[37]

All service members who went through interviews remember the barrage of questions. Richard Gration, investigated in early 1982, wrote notes after his interview that described the RAAF Police tactics. For instance, the flight sergeant advised 'it was in my interst [*sic*] to answer all their questions as no answer would imply that I was guilty … He also said that "this was my opportunity to put my side of the story across". This was not recorded on the record of interview. This was the first time I had received a caution.' Even though the policy on homosexuality indicated that Richard had the right to request an officer be present, the flight sergeant 'emphasised that the only reason I would want an Officer would be if I felt physically threatened. He also said that if I asked for an Officer I would have to wait until the Officer was brought in.'[38]

Eliciting a confession was not enough. Notwithstanding the 1974 instruction intended to limit the line of questioning, police regularly asked for specific details about sexual practices and partners. Kyle Duffield's RAAF Police interview transcript from 1989 includes the questions: 'Kyle, in what form was the relationship with X?' 'Kyle, you stated that you do not partake in anal sex. Could you tell me what form you do partake in?' 'Kyle, have you ever partaken in mutual masturbation?'[39] A 1982 RAAF Police interview transcript details 'Ryan's' vivid descriptions of his sexual experiences with both ADF personnel and civilians. Ryan's transcript also includes the questions: 'Did you enjoy fondling X's penis and performing fellatio on him?' 'Have you ever had sexual intercourse with a female?' 'Did you enjoy these experiences with females?' 'Did you ejaculate whilst having these experiences with females?'[40]

Some accused answered with short yes or no responses; others limited their replies to vague sexual details, trying to downplay their homosexual activities or, as in the case of Richard Gration, explaining them as drunken or passive encounters.[41] Others, like Ryan's interview transcript, went into great depth describing their sexual performance. 'Peter' recalls reading one interview transcript in the RAN legal office with a cheeky answer:

> The record of the interview goes something like this: 'And Seaman Jones, did you have sexual intercourse with Seaman Bloggs?' or whatever. And Seaman Jones says, 'Huh?' 'And, did you commit an act of fellatio with Seaman Bloggs?' 'Um, ah …' 'Did you have sexual—did you commit an act of fellatio?' 'Oh, sort of.' So, the naval policeman says to Seaman Jones, 'What do you mean

"sort of"?' And Seaman Jones says, 'Well, I tried, but his cock was too big.'[42]

Lieutenant Alix Blundell's 1988 interview with the SIB was a harrowing experience that traumatised her for life. At the time she was experiencing chronic back pain from an injury she sustained during an Army fitness test, as well as a blown knee. Blundell requested a commanding officer, but the SIB interrogators denied the request and threatened her with dishonourable discharge if she failed to cooperate during the interview. Interrogators denied her legal representation. Only after some discussion with her commanding officer did they permit his attendance but only as an observer. The interrogation went on for hours, with lines of questioning and verbal abuse similar to other examples. During the interrogation Blundell's legal officer and psychologist attempted to gain entry but were denied. The interview caused not only mental duress but also exacerbated Blundell's physical condition. As she explains, 'I was in terrible pain, and they wouldn't let me go to the toilet. They wouldn't give me water. They wouldn't allow me pain relief.' One of the interrogators even proudly declared that she was a born-again Christian and was doing this to serve her god. Blundell did not crack during the interview and never confessed to being a lesbian.[43]

The consequences of Blundell's interview were far-reaching. Just over a month later she suffered a mental breakdown and had to be hospitalised. After her release from hospital, SIB continued to pressure Blundell to resign her commission, even though she never admitted to being a lesbian. Blundell refused, so her security clearance was downgraded and she could not perform her job as an education officer. She lodged a redress of grievance against the SIB. The process of investigating her redress and her appeals up the chain of command took more than a year. At last Blundell had one minor victory: the Army Chief of General Staff agreed with the redress of grievance that the SIB had mistreated her and not followed appropriate investigation procedures. Unfortunately, in the interim, after continuing threats of dishonourable discharge, Blundell had requested her own discharge in November 1989. The Chief of General Staff made it clear that she could not re-enlist in the Army because of suspicions over her sexuality. Blundell was never able to return to full-time work because of enduring physical and mental health problems.[44]

For the next fifteen years Alix Blundell fought for justice anywhere she could: the Department of Defence, Department of Veterans' Affairs and through representations to parliamentarians. It was only when the Commonwealth Government set up the Defence Abuse Response Taskforce (DART) in 2011 that Blundell was finally treated with respect. In 2014 DART determined that Blundell was a victim of abuse and awarded her financial compensation and a restorative engagement package that included a personal apology from the then Chief of Army, now CDF, General Angus Campbell.[45] Meeting Blundell and listening to her story really touched Campbell, who subsequently requested that Blundell be treated as a medical discharge. In addition to a typed letter of apology, he handwrote a personal letter that said:

> You were treated in a degrading, damaging and wholly unlawful manner, for which I am deeply ashamed. It is clear to me that you were powerless to act and that your inquisitors were acting beyond any ethical or professional basis. I hope, in that light, you might allow me and more generally the Australian Army to bear the shame and guilt of our misconduct. Your service is appreciated and commendable, and I hope you can recognise this truth. Please be assured I am determined that the Army learns from its past by owning and dealing with the truth of that past.[46]

It took another five years of fighting before Blundell received a Commonwealth Superannuation Corporation statement of Medical Discharge, but only at Class C 10 per cent injury. After a reconsideration request, in February 2020 Blundell received a full Medical Retirement pension of Class A (incapacity of 60 per cent or more), with the date of Medical Retirement being 20 November 1989.[47] It took more than twenty years, but at last the Commonwealth has acknowledged Blundell's physical and mental injuries, then and now.

The witch-hunt trope

Investigations into homosexuality rarely happened in isolation. Service police pressured suspects to name other LGB people in the ADF. For instance, Lyndal Green recalls her interactions with SIB interrogators: 'You know, "Make it easy on yourself; dob in a mate." "Let someone else go to the wolves. We'll go easy on you. Dob in three mates." She fiercely refused: "No, I'm not having a bar of it."'[48] Many others also resisted naming LGB members, either because they did not know any or because they wanted to protect their friends. Susie Struth, dismissed from the Army in 1977, recalls the SIB agent presenting her with the nominal roll and asking her to mark the lesbians. She refused by retorting, 'It would be quicker for me to tick off those that aren't.'[49] Leon Fry worked as a clerk in RAAF administration and saw many police investigation files cross his desk. He recalls: 'Interestingly, nobody really spilt their guts on other people. [It] wasn't like if people were caught with drugs and things like that. Drugs was a different story altogether and people would spill, "Oh, so-and-so and so-and-so." But when somebody was gay and got caught, they didn't seem to spill the beans on other people.'[50]

Given how investigations spread across bases and social networks, clearly there were LGB service members who did name others. For instance, Shane Duniam was one of five airmen discharged in late 1981. Authorities had been investigating them for months, with one police interview with an airwoman confirming their homosexuality and alleging (falsely) that they used to host parties to watch pornography.[51] The Melbourne tabloid newspaper *Truth* even reported their discharges under the headline '5 Sacked in Camp Probes'. The RAAF representative commented, 'It seems that once we catch one ring of homosexuals, other similar incidents are reported.'[52] Sandy Ashton recalls a period at RAAF Base Edinburgh in the mid-1980s when the police kept summoning members for hours on end, seeking the names of gays and lesbians. She states: 'I know people disappeared [were discharged], but I can't remember who they were.'[53] These series of investigations to uncover LGB members became colloquially known as witch-hunts.

By the 1980s the term 'witch-hunt' had well and truly entered the Australian lexicon to describe the investigations targeting LGB Defence members. Indeed, the RAAF representative in that same 1981 *Truth* article declared, 'We have not been involved in a "witch hunt of homosexuals", it is purely coincidental that there has been a spate of incidents.'[54] The ADF continuously rejected the witch-hunt trope, with the Standard Explanatory Position accompanying DI(G) PERS 15-3 even stating: 'Suggestions that the instruction may result in a "witch-hunt" to unearth homosexuals in the ADF are without foundation ... There is no intention to invade members' bedrooms to seek out homosexuals.'[55] Although the ADF was denying that witch-hunts happened, its use of the term was an implicit acknowledgement of the importance LGB members and other critics placed on the expression.

Sometimes LGB members had advance warning that a witch-hunt was coming. They may have heard about investigations on other bases or were forewarned by friends. WRAN Ronni Spencer remembers receiving a warning while she was serving in Darwin in 1982:

> We'd been there about a year, and we then heard that there was a witch-hunt happening, so that was very scary. They were saying that they were going to interview people and come and inspect your house and blah, blah, blah. We got quite worried, so we collected all the things that they might find as evidence, like letters and things like that, and we had a little bonfire and burnt all our cards and letters so that there was nothing they could pin down on us. But nothing ever happened, so we ... survived that hunt.[56]

Both men and women recall witch-hunts, but oral histories suggest that witch-hunts increased around the mid-1980s and targeted servicewomen more than men. Former soldier Sandra McInerney believes authorities targeted lesbians more than gay men because 'I think maybe people, other straight women, were threatened, maybe'.[57] This was certainly the case in a prominent 1987 witch-hunt that multiple servicewomen remember. It began at an interservices hockey competition when a lesbian soldier hit on a straight woman. The straight servicewoman reported the incident, and from there the investigation spread to investigating most of the women who were present at the tournament. Both Gen Ford (Army) and Kate Carlisle (then Navy) remember this investigation, with Carlisle describing it as a 'massive witch-hunt. Massive witch-hunt. God. Girls in Brisbane, Melbourne, Sydney. Yeah, it was just … it was bloody ballistic.'[58]

Although witch-hunts seemed more prevalent among women, the AIDS epidemic also launched and exacerbated witch-hunts targeting gay men beginning in the mid-1980s. Stuart Martin was a nurse in the RAAF, and he recalls: 'So we started to get people coming in who were getting sick and having blood tests and who'd been diagnosed. And then suddenly there would be a witch-hunt. The military police would want to know everyone they'd had contact with.'[59] Reinforcing Martin's memory, the April 1991 *OutRage* article said:

> You read periodically about incidents such as one in Darwin a couple of years ago, where they found one HIV positive person and uncovered a network up there. There was a big outrage in the Parliament, questions asked in the Senate about what was going on in Darwin. In fact the military police there, instead of treating HIV as a medical issue, had treated it as a security problem, with no compassion or understanding. The HIV positive people, who had shown up in routine testing, were interviewed by service police, not by doctors, in an attempt to establish what other gay people they knew.[60]

More information about the ADF's policies towards HIV and AIDS is available in chapter 5, but as this example shows, a positive diagnosis was regularly treated as a proxy for finding gay men.

Discharges

The 1974 policy on homosexuality had some leeway for the ADF to retain members if the investigators determined that the accused was not a 'confirmed homosexual'. This distinction grew out of the Navy's pre-1974 policy on 'Abnormal Sexual Behaviour' and drew on pre-1973 psychoanalytical understandings of homosexuality. Developmental psychologists since the 1930s argued that homosexuality was a childlike state of development which all men (and the focus was on men) went through, but which they should outgrow as they matured sexually. Homosexuality was the result of either men who did not develop properly, or of men who 'reverted' to homosexual acts when there was not an adequate heterosexual outlet. Psychology studies, and even the Wolfenden Report (1957), which recommended decriminalisation of consenting adult homosexual acts in the United Kingdom, explicitly mentioned the military as one site where men who would otherwise be heterosexual may revert to performing homosexual acts.[61] Therefore, the idea of a confirmed homosexual essentially was someone a psychiatrist certified as not having developed 'properly', whereas an experimenter was merely a heterosexual who (supposedly) erred and reverted briefly to a childlike stage of sexual development.

Although new psychological and psychiatric research post-1973 (when the American Psychiatry Association removed homosexuality from its *Diagnostic and Statistical Manual of Mental Disorders*) shifted away from these developmental explanations for homosexuality, the Defence policy retained the outdated distinction. The 1974 instructions were vague about how to differentiate a confirmed versus unconfirmed homosexual: 'account is taken of such factors as:

(a) was the incident an isolated one?
(b) can the incident be ascribed to adolescent experimentation?
(c) is there a psychological or psychiatric assessment that the incident was non-typical and unlikely to be repeated?
(d) how widely known is the incident?'[62]

Army policy in 1982 elaborated slightly on these questions, such as asking whether the incident caused a public scandal, departed from the member's 'usual and customary behaviour' or could be ascribed to intoxication.[63] Even this was not enough to clarify how to assess whether someone were a confirmed homosexual—a point made by Army psychologists in 1985.[64] During a 1991 review of the policy, a lieutenant-colonel in the Directorate of Personnel Employment queried:

> Is it meant to imply that a soldier can commit a homosexual act (to what degree) and not be a confirmed homosexual? Is a CO [commanding officer] qualified to be the judge of such human

> behaviour and let the soldier off with a warning? If strict guidance is not stated, there is plenty of scope for the policy to be inconsistently applied (although one would hope not).[65]

Another officer commented: 'Making allowances for "experimenters" is a luxury that the military can ill afford. The damage that the retention of such a person is likely to cause to unit cohesiveness and morale far out weighs the benefits associated with saving one soldier.'[66]

In practice, the nebulous distinction between confirmed homosexuals versus others did not matter because service police almost always equated homosexual behaviour with homosexual identity. The 1991 letter from the anonymous ex-RAAF policeman to *OutRage* stated: 'There is no such thing as giving a "warning" to one-off sexual transgressors, as Govt psychologists have advised the ADF that a person does not dabble occasionally, infrequently, or by accident in sexually deviant behaviour. The ADF cannot take the risk that the person might re-offend "morally".'[67] From 1985, DI(G) PERS 15-3 used the language of targeting 'homosexual behaviour' rather than whether someone identified as lesbian, gay or bisexual. Such an approach also meant that bisexuals, who were only ever mentioned once in an unpublished draft of DI(G) PERS 15-3,[68] were always treated as if they were confirmed homosexuals.

Only two participants in this project successfully argued that their homosexual behaviour was experimentation and were let off with warnings. One was Richard Gration, whose powerful connections, as chapter 4 explains, gave him the opportunity to challenge the RAAF Police. The other was Darran Clark, who was investigated in 1987 after being reported to the Naval Police by an ex-lover. Clark had something in common with Richard Gration: a powerful ally. His ship's captain, David Farthing, had a law degree and had even worked as a barrister. Clark recalls: 'What he said was that, "We're putting this down to experimentation and there will be certain things you would have to do if you're going to be retained." And I was like, "OK." And I wasn't happy about that but, you know, I also didn't want to leave the Navy.'[69]

There are also some circumstances where, for reasons unknown, the services did not pursue the discharge of an LGB member. One cable from September 1982 notes that 23 Field Regiment was allowing a soldier to retain his role after a 'minor' incident, although the soldier was warned that 'any further incidents will lead to his immediate discharge'.[70] 'Simon' was a cook in the Navy who, in 1990, was arrested at a department store in Melbourne for attempting to solicit sex in a public toilet. Victoria Police contacted the Naval Police, who escorted Simon back to HMAS Cerberus. Naval authorities directed Simon to see both a psychologist and a psychiatrist, who wrote a report. Simon does not know what was in the report, but he never was prosecuted for homosexuality. Did the report suggest that Simon was not a confirmed homosexual, or was there some other aspect of Simon's case that meant he was permitted to stay? What makes Simon's retention even more surprising is that he pleaded guilty in a Melbourne magistrate's court to the summary offence of public indecency. A reporter from *Truth* was at that hearing, and the headline the next day was something to the effect of 'Sailor takes pot shot in toilet'. Still, Naval Police never went after Simon. He endured some bullying and harassment, which mostly ended when he was deployed on a ship a year later and was subsequently posted to HMAS Stirling in Perth.[71]

In the majority of cases, however, the investigation ended in a discharge. Not surprisingly, most LGB people elected to take the honourable discharge at their own request because a dishonourable discharge would form part of their record and adversely affect their future employment prospects. Most discharges were relatively straightforward, processed within a matter of days. Janet Carter initially tried to fight against being forced out. Ultimately, however, it was made clear to her that 'if we can't make you leave, we're going to actively find ways to make your life so uncomfortable that you won't want to stay'. Five minutes after she had told her lieutenant colonel that she would resign, she walked back to her room where she found that 'they had cleaned out my room. They had taken everything. All my uniforms and stuff, and they put my bag on my bed, my suitcase on the bed with the lid open, waiting.'[72]

In cases where someone continued to deny their homosexuality—such as Alix Blundell or Yvonne Sillett—the process could be drawn out. In those instances, the police reduced or denied their security clearance, thus making the service member unable to perform their job and often compelling them to request their own discharge anyway. In some instances when an LGB member was close enough to their end of service, commanding officers agreed to let them quietly serve it out rather than force an early discharge.[73] In August 1981, a RAAF group captain praised the work of a gay airman recently investigated and recommended that he 'be retained in the RAAF at least until the expiration of his present engagement. This recommendation has been carefully considered and is made with … the realization that homosexuals are in general regarded as socially abnormal and are a potential security risk.'[74] The outcome of that petition is unknown, but it shows that commanding officers were willing to advocate for and/or tolerate the retention of some LGB members.

Colin Edwards's story of investigation and discharge is certainly one of the most dramatic recorded for this project. In 1980, a man in a RAN officer's uniform approached Edwards at a Kings Cross bar in Sydney and warned him that he was under investigation for homosexuality. The man claimed to be a Navy Reserve lawyer and said he was there to defend Edwards. Twice Edwards went to the man's residence, where the man advised him that he was one of several gay men under investigation. The man tried to record a statement from Edwards, but Edwards would not comment because something felt wrong. Edwards voluntarily turned himself in to Naval Police and underwent the same long, detailed interview process as other LGB suspects. Edwards's interview took a different twist at the end. He explains:

> Then they pulled out a picture, a photograph, and they showed me a photograph and it was this fellow in his officer's uniform. They said, 'We know this man. We've been chasing this man for almost three years. This is a man that is guilty of impersonating a naval officer, and he's a part of a larger group of guys that attempt to lure young men into sexually compromising positions, for the purpose of blackmail.'[75]

Naval Police persuaded Edwards to participate in a sting to catch the fraudulent reserve lawyer. They arranged for him to meet the man in a bar three more times, and the plan was for police to arrest him at the final meeting. The man did not show up, though; when Edwards rang him from the venue, the man told Edwards to go to hell and hung up. The entire operation was a bust.[76] Naval Police were angry, and someone had to pay after all they invested in catching the fraud.

Shortly after the failed operation, Colin Edwards deployed on HMAS *Melbourne* in what was a memorable tour in which the ship famously rescued ninety-nine Vietnamese refugees from a sinking boat.[77] For Edwards, there was a second dramatic moment when he was unexpectedly hauled before the captain's table. The officers advised Edwards that he was being charged under the policy against homosexuality, and he had two choices: sign a document requesting his own discharge, or else be dishonourably discharged 'services no longer required'. Like so many others, Edwards signed the form.

Within four hours, HMAS *Supply* was alongside *Melbourne* and Edwards was tied to a rope and sent across to *Supply* to be returned immediately to Australia. This very public action was designed to humiliate and terrify him.[78] It also terrified other gay service members serving on *Melbourne*. Kenton McKeown, then a young sailor who did not yet realise his sexuality, vividly recalls what happened: 'Colin was then kicked off the ship and I was told that's what happens to faggots and if I ever find one, make sure everyone knows.'[79]

Dramatic though Colin Edwards's experience was, at least his aligned with the policy on homosexuality, and he ended up with an honourable discharge. Others had worse outcomes when authorities did not properly follow the policy. 'Peter' recalls a case in 1976 when two sailors were caught and given different punishments. The younger, junior sailor had a reduction in his security clearance and was retained by the RAN. The older, more senior sailor had his security clearance revoked, which meant he had to be discharged. When Peter queried the unequal punishments, the response was that the older sailor was married and therefore more vulnerable to blackmail.[80]

In a more extreme example of inequality and broken rules, 'Bruce' and another airman were caught naked in bed and investigated in 1990. The other airman was from a different mustering and, in accordance with the policy, given the opportunity to request his own honourable discharge. Bruce was not; he was simply sent to Darwin where he obtained a dishonourable discharge. He then went to the beat at Mindil Beach:

> Stripped off my uniform and left it sprawled around there because I didn't give a shit. This was just, I guess, one way of dealing with it … and just left them [uniform clothes] lying there. And obviously changed and … Fuck, I think I cried because I had no idea what I was going to do. And very much alone, didn't want to talk to anybody.[81]

Bruce's case highlights not only the sense of powerlessness that most LGB service members felt but also the sheer psychological trauma that investigations and discharges often inflicted.

Cost of investigations

Police investigations—whether they ended in discharge or not—took an enormous toll on the accused. Those who discharged faced financial and career uncertainty, not to mention the complication of explaining to their family and friends why they left the ADF. Some were open about the reasons, but most lied with vague statements about not being satisfied with the job any more. Some kept the secret of their discharge for years. Danny Liversidge announced in an emotional Facebook post only in 2016 that he had been discharged from the RAAF in 1991 for being gay.[82]

For many, military service was an all-consuming role and being in the ADF was a core marker of their identity. After Colin Edwards's dramatic discharge from the Navy in 1981, he had no idea what to do with his life. He recalls:

> I was totally destroyed. The only job I ever wanted to do, the only thing I ever wanted to do, they took away from me. You know, on one hand they said that I wasn't good enough to lay down my life

for my country, or that's the way that I'd perceived it, you know? As a result of that I was bewildered, hurt, emotionally distraught and pretty confused.[83]

Edwards's statement has echoes among others dismissed from the ADF. These people needed to find a way to rebuild their sense of self, and just severing the link with the ADF was often a challenge. Army Captain Gen Ford, compelled to resign in 1988, started her own recruiting business, which specialised in finding positions for ex-ADF members. She says of this job: 'So it sort of kept this link in a weird way, and like at the time I probably didn't quite realise it, but when I look back at it I go, "Yeah, that was my way of staying in touch with the Army."'[84]

Many LGB service members facing investigations or discharge considered suicide. 'Mark', discharged from the Army in 1988, describes contemplating suicide 'on many occasions' and extreme feelings of anger. At one stage he went into such a rage that he punched a man who was taunting him and a friend outside a Melbourne gay bar.[85] In another example, RAAF Police investigated 'Nathan' for drug use in 1989. When searching his bedroom, they came across gay magazines and a bottle of amyl nitrate, so Nathan confessed to being gay. He recalls: 'I was, I suppose, feeling sort of suicidal that I would actually, like—well, I'll admit it [being gay], and then sort of drive my car into something, into a truck, and I think that's what was going through my head at the time that this is OK, sort of a relief it's over.' The police referred Nathan to a psychologist, who was very supportive and worked with him on relaxation therapies. For reasons unknown, the police never prosecuted Nathan for homosexuality, and he continued to serve for another eighteen months.[86]

Others not only contemplated suicide but actually attempted or succeeded. One suicide survivor is Richard Williams, who was discharged from the Army for homosexuality in 1974. Williams drove from Townsville to Sydney, then across Australia to his parents' home in Perth. One night his mother asked him the real reason he left the Army, and he showed her a copy of his police interview transcript. Her response was: 'It doesn't matter to me. You're my son and I love you very dearly and if you find another man that you would like to partner with, you'll be welcome to bring him home at any time and we'll make him feel very welcome.'

Williams burst into tears and, notwithstanding his mother's loving support, decided to end his life. He went to the medicine cabinet and took a deadly cocktail of drugs and alcohol. He barricaded his bed against the door and, while writing a series of farewell letters, collapsed. Unknown to Williams, his mother was still awake. Not able to enter his room owing to the barricade, she went outside, smashed the window, climbed into his bedroom and performed CPR on him until the ambulance arrived and transported him to Royal Perth Hospital. Two days later, Williams awoke from his coma in hospital.[87]

Other LGB service members did not survive their suicide attempts. An Army statistical breakdown of soldiers investigated for homosexuality between 1987 and 1992 noted that one member suicided.[88] Former WRAAC Jennifer Jefferies recalls: 'A few girls in our squad were gay, two of them attempted suicide, one of them took her life finally, after she got out of the Army.'[89] Several other WRAACs recall young women who suicided or attempted suicide early in their careers, although they are unclear about whether they did so because of their sexuality.[90] 'Brett' remembers a gay Army mate who suicided in Townsville in the 1980s, shocking the other members of their squadron.[91] Neil Murray recalls at least two cases of airmen who suicided because of the strain of hiding their homosexuality.[92]

Suicides affected not only family and friends but also those who served alongside the deceased. Vice Admiral Ray Griggs—who served as Chief of Navy (2011–14) and Vice Chief of the Defence Force (2014–18)—lost a friend to suicide in the mid-1980s, and it had a profound transformative effect on him. He says:

> I think the most telling thing for me was when a colleague of mine, who I won't name, was worried. He thought he was going to be outed. He was doing a course out at HMAS Watson at the time and killed himself. That, I think looking back, that's probably the moment for me that [pause] I really thought that this [discrimination] was wrong.[93]

We will never know how many LGBTI service members suicided either when they were caught or feared being caught, or because they were struggling more generally with their sexuality, gender identity or intersex variations.

Statistics on investigations and discharges

Some quantitative data about the numbers of investigations and discharged LGB service personnel is revealing about whom the service police targeted. In early 2018, under a Freedom of Information request, the ADF Provost Marshal's office released data confirming that police in the three services investigated at least 489 men and 165 women for homosexuality between 1967 and 1992.[94] This tally is not complete; it is based on records that the Provost Marshal's office compiled while preparing documents for the Royal Commission into Institutional Responses to Child Sexual Abuse.[95] Several of the ex-service members interviewed for this project requested copies of their own police records, but they could not be located among the 654 case files. Moreover, the Provost Marshal's office advised that many old RAAF Police files were destroyed in 1993. Notwithstanding these limitations, the incomplete tally reveals a consistent stream of investigations, especially after 1973. It also shows that women were disproportionately targeted—being 25 per cent of the investigations despite the fact that they were less than 10 per cent of the ADF for most of this era.[96]

More useful data comes from a question on notice at a Senate Estimates Committee hearing in September 1992, which queried the number of people discharged for homosexuality during the five-year period 1987–92. The numbers in the internal ADF documents do not add up perfectly, but the figures published in Senate Estimates indicate that thirty-two RAAF, seventeen Army and twenty-four Navy members accused of homosexual conduct requested their own honourable discharges; there were also thirteen RAAF, five Army and three Navy dishonourable discharges (see tables 2, 3 and 4).[97]

Table 2: Homosexuals dishonourably discharged as reported to Senate Estimates Committee, 1987–92

Financial year	Navy	Army	Air Force	Total
1987/88	nil	nil	2	2
1988/89	1	nil	2	3
1989/90	2	2	6	10
1990/91	nil	2	3	5
1991/92	nil	1	nil	1
Total	**3**	**5**	**13**	**21**

Source: Australian Senate, Estimates Committee B, Report to the Senate, '1992–93 Senate Estimates Hearing Committee B—23 September 1992 Questions on Notice', pp. 144–5

Table 3: Homosexuals who requested their own discharge as reported to Senate Estimates Committee, 1987–92

Financial year	Navy	Army	Air Force	Total
1987/88	11	4	16	31
1988/89	4	1	7	12
1989/90	1	5	4	10
1990/91	3	6	5	14
1991/92	5	1	nil	6
Total	**24**	**17**	**32**	**73**

Source: Australian Senate, Estimates Committee B, Report to the Senate, '1992–93 Senate Estimates Hearing Committee B—23 September 1992 Questions on Notice', pp. 144–5

Table 4: Aggregate internal figures for report to Senate Estimates Committee, 1987–92

Service	Honourable discharges	Dishonourable discharges	Investigations: male/female	Officers
Army	17	5	25/31 (55% F)	1 or 2
RAAF	32	13	unknown	3
Navy	24	3	22/5 (18.5% F)	unknown
Total	**73**	**21**	—	—

Source: Data compiled from NAA A6721, 1985/18156 part 4 and part 5

It is difficult to interpret this quantitative data. Does the relatively small number of investigations and discharges across the services suggest that perhaps investigations were not so widespread, or does it suggest that accused individuals were successful at not naming others? Bear in mind that the number of cases for 1992 is lower in all three services because the LGB ban was under review and the CDF placed a moratorium on investigating cases of homosexuality from December 1991 to June 1992.[98] The data does reinforce some of the oral histories. For instance, in the RAAF and RAN, there was a spike in offenders reported in 1987–88—the period coinciding with the purported witch-hunts.

The Navy and Army statistics include gender breakdowns. In the Navy, only five of the twenty-seven discharges (or 18.5 per cent) were women. Although this figure seems high, actually women were approximately 12 per cent of the Navy in 1990, so it is not excessively disproportionate (particularly given the small sample). In the Army, however, thirty-one out of fifty-six investigations (55 per cent) were women; of the discharges (both honourable and dishonourable), sixteen (67 per cent) were women and eight (33 per cent) were men. In every reported year except 1987, there were more investigations of women than men. Given that women made up 8.6 per cent of the Army in 1990, this is clearly a sign that the SIB were targeting women/lesbians more than gay men.[99] Given the longstanding stereotypes, restrictions and anxieties about lesbianism outlined in chapter 2, it is not surprising that authorities were still more prone to police women's sexuality. Indeed, it was sometimes said that if a servicewoman was neither married nor sleeping around with servicemen, she must be a lesbian.[100]

There is another intriguing observation in the quantitative data: officers versus other ranks. The Navy data does not include this breakdown, but both the RAAF and Army statistics clearly show that far fewer officers were discharged for homosexual acts than other ranks. The RAAF charged only two officers and one officer cadet, while the Army only charged one or two officers (the internal documents conflict).[101] There are several possible reasons for this disparity. Because officers were in a higher position with more autonomy, there would be less scope for police to investigate officers unless they had more probable cause. There also might have been more secrecy about the identities of gay officers. One 1989 Army report about an investigation that uncovered nine gay other ranks indicated: 'Several officers from [redacted] were also allegedly involved in homosexual conduct. To date, however, it has not been possible to identify any of these officers.'[102]

The oral histories also suggest that the ADF was more hesitant to investigate, let alone discharge, officers over homosexuality. Neil Murray recalls visiting beats, including gay beaches in Adelaide or Canberra, where he regularly saw other officers such as a group captain who was Director of Personnel Officers in the Air Force Office—the very man responsible for kicking people out for homosexuality. Murray also has insight because he worked for most of his RAAF career from 1965 to 1984 as an adjutant: the person assisting a base commanding officer with administrative tasks, paperwork and discipline. At numerous bases Murray was responsible for processing the paperwork relating to male homosexual discharges (not women). When asked about officers, Murray responded:

> Officers' cases did come but there was no occasion where an officer was ever discharged. It was more pick on the lowest ranks ... Well, put it this way, they would be formally warned, cautioned and given a reprimand or a warning and told, 'Listen, one step wrong again, you're gone.' So, it put the fear into them and put the fear into me. Yes, there was a rule for one and a rule for another.[103]

Of course, officers were not completely immune from investigations and discharge, but Neil Murray's observation and the quantitative data imply that there was a higher threshold before police could target and investigate them.

Gay service life under the ban

In one sense the service police investigations were effective because they resulted in the discharge of hundreds of LGB service members. The investigations and witch-hunts frightened LGB members to keep their sexuality secret, conscious that they could be exposed with one false move, the wrong photograph, or by associating with the wrong people. Yet LGB people also found strategies to cope, ranging from staying deep in the closet through to forging relationships and social networks to protect each other. Chapter 2 explains the lesbian subcultures in the women's services. These continued into the 1980s, and women continued to live, socialise and protect each other after the disbandment of the WRAAF, WRANS and WRAAC in 1977, 1984 and 1985 respectively. Women had to contend with the dual pressures of both homophobia and sexism in this era, and support from colleagues was critical.

Gay and bisexual men had different experiences from women as they were rarely surrounded by so many others who were attracted to the same sex. Some gay and bisexual men married and had children. Some married before they realised they were gay, while others did so either to try to change their sexuality or to deflect suspicion. Most of the married gay or bisexual men interviewed for this project were unfaithful to their wives. Some occasionally visited a beat; others had long-term affairs with other men. Michael Dinnison, who was an Army officer who reached the rank of lieutenant colonel, was so smart as to plan visits to sex clubs and beats or with regular or semi-regular partners around his wife and children's schedules.[104]

A marriage or relationship with a woman was an important shield (or 'beard', to use a gay colloquialism) that provided a veneer of protection for many gay men. Indeed, as Brian Lawrie explains, a single male in his thirties could arouse suspicion among service police.[105] While many gay servicemen married or dated straight women who unsuspectingly thought the men were straight, a few formed relationships of convenience with lesbian servicewomen. Airwoman Sandy Ashton even recalls an evening when her drunk gay friend Stuie Watson proposed to her. She explains, 'It felt like that would give us a bit more comfort in knowing that they won't come after us, but we could still continue on with our lives, our gay lives. But I said no.'[106] 'Bruce' did not have any fake relationships, but several lesbian airwomen spread rumours that they had slept with him. This effectively deflected attention from themselves and from Bruce, who developed a reputation as a ladies' man.[107]

Both married and single gay or bisexual men searched for sexual outlets, and during the 1980s the capital cities all had a thriving network of sex clubs, saunas and beats. Visiting sex clubs or saunas ran the risk of being followed by service police, but still gay/bisexual men were willing to take that risk. RAAF pilot 'Bill' recalls: 'I'd been down to a gay sauna in Melbourne, and then a week or so later one of the colleagues who was working at a managerial level made a joke about jumping into a sauna with your mates, and directed at me, and there was no logical other explanation for that.'[108] Visiting gay bars ran the risk of running into undercover service police. RAAF officer Brian Lawrie took steps to avoid detection, as well as surprise encounters at gay bars:

> Of course you become quite paranoid, and so I used to be. I'd wait until dark, and I'd park about a block away from the sauna, and that sort of behaviour. You had to be very careful … then, one night, I remember, one of the gay hotels there, a gay pub, it was,

> and one of my corporals … just looked at me, and was shocked, and he said, 'What are you doing here, Sarge?' And I said, 'Well, same as what you're doing here.' But we never spoke, or the subject was right off the books.[109]

Several ex-servicemen recall running into others at beats, including seeing some very high-ranking officers.[110] Their oral histories indicate an unspoken code of never speaking about these sightings with each other. Visiting beats was dangerous business. In the 1980s especially, straight men would go to known beats to bash and sometimes even kill gay men. This violence, and possible police involvement and cover-ups, has recently become the subject of New South Wales coronial inquests and a parliamentary inquiry.[111] Navy officer 'Kevin' nearly became a victim of one such hate crime while visiting a beat in Manly, northern Sydney, in the 1980s. He escaped attack narrowly when he ran away from a man who had entrapped him.[112]

Beats and saunas presented opportunities for sexual release, but they were not usually sites of emotional intimacy. Whereas chapter 2 showed that intimate relationships were common for lesbian and bisexual women, this was not the case for gay men. Partly this was a symptom of gay culture more widely in the 1970s and 1980s, where the commercial scene promoted physical appearance and short sexual encounters.[113] Many did not desire a relationship because they feared they might be discovered. Others, like Army officer Brian Lawrie, found it hard to articulate why they shied away from relationships: 'No, I didn't start going out into the gay scene with civilians as such. No, you might meet somebody and go to his place, or he'd come to your place or something, but I didn't take on a permanent relationship, no, because—I suppose some service blokes do, but I didn't.'[114]

Military service compounded the challenges all same-sex couples faced. Across Australia LGB people remained closeted during this era for fear of ostracism, losing family or being dismissed from their employment (as there were no anti-discrimination protections for gays and lesbians except in New South Wales from November 1982, South Australia from 1984 and the Australian Capital Territory from 1991). Military service had two extra burdens: the ever-watchful eye of the service police, and the demands military service places on any relationship because of postings and deployments. When Craig Cahill first joined the RAAF in 1985 he was dating a guy named Ian. To disguise their relationship, he used the longstanding practice of writing letters addressed to a female name: 'Jan'. After recruit training, Cahill was posted to Sale in eastern Victoria, and his relationship ended shortly thereafter.[115] Airman Leon Fry recalls that secrecy prematurely ended one relationship: 'He [partner] kept on saying to me, "You're not being honest about everything about you", he said, "so I don't trust you." Well, I wasn't either, because I didn't trust him.'[116] For those who did manage to maintain long-term relationships in secret, there were challenging moments for which they had little support. 'Kevin' recalls that when he deployed for the First Gulf War in 1990:

> I had to go and say goodbye to Craig outside the airport in the car, and it was very emotional; he was crying and I was crying, because I was going to go away to the Gulf War. So that was quite an emotional time. But during the three months we were on the ship, he wrote to me and I wrote to him, but I wrote to him in non-sexual terms. So we write all these beautiful letters, but it would be 'Dear Lover'; it wouldn't be 'Dear Craig', because I was afraid somebody would find them. And I still have all those letters, actually, from that time.[117]

Many men and women use the expression 'double life' to describe their service years. They had different ways of coping with the stress of hiding their sexuality and social lives, knowing that one wrong word or trusting the wrong person could lead to their discharge. Delia Quigley explains:

> You'd start to compartmentalise your life and hide the private stuff about what you did on the weekend, who you went out with, who your friends were. You kept it separate, a lot, a lot more separate from the workplace than what you would … do if you were just hanging out with all the heteros and were one of them … It was about preserving yourself and keeping your job and not being placed in that position where you were going to be dragged into being interrogated, and face losing your job.[118]

Some could not cope with the secrecy, as evidenced by those who voluntarily outed themselves and by the tragic suicides. Others had no mental health problems because they were already accustomed to living a double life. As Flying Officer 'Archibald' points out, 'I think I was able to departmentalise it, because I think up until that point I had been doing that unconsciously or subconsciously from probably puberty onwards.'[119]

Some men not only visited gay bars but also became involved in the emerging gay communities where they were posted. For instance, RAAF pilot Richard Gration briefly participated in a group for young gay men under twenty-six called the Sydney Gay Youth Group.[120] While based in Darwin from 1988 to 1990, RAAF Corporal Craig Cahill worked for the Gaywaves program on the university radio station and volunteered for the Crisis Line counselling service.[121] A mate grabbed Army Captain Tony McLeod from the sidelines to march in the 1988 Sydney Gay and Lesbian Mardi Gras Parade (where he was spotted and brought to the SIB's attention).[122] Able Seaman Colin Edwards was friends with many drag queens, who made him march in the 1980 Sydney Gay and Lesbian Mardi Gras in disguise. He explains:

> I'm not a drag person. So, they put me in a clown costume. Well, it wasn't a full one. I was in jeans and I had this funny white shirt with the big fucking bow tie, and they painted a white face on me, and then, a bit like clown makeup, if that makes sense. Then I went with them on that.[123]

Some men were quite bold in their behaviour. Keith Jeffers recalls one amusing encounter at the canteen on RAAF Base Wagga: 'I bought some Vaseline and the woman said, "What do you want to use that for?" I said, "My lips." She replied, "That's fine then." I suddenly realised that I was being questioned about why I needed to buy Vaseline.'[124] In the late 1980s, Tony McLeod sometimes even brought men he met on Oxford Street, the heart of Sydney's gay scene, back to Victoria Barracks down the road in Paddington. He explains:

> There were people who lived in [the barracks], but they would maybe live in there Monday to Friday and then they would go home for the weekend, so I would kind of be an alternate shift. So there was really nobody around to see. Yes, there was a chance [of being caught], but that kind of adds to the edge of making the whole thing exciting and the boys loved it. They'd say, 'Oh, going to the Army barracks', so it was fantastic.[125]

Sex on base could be risky, as attested by so many LGB people who were caught. By the 1980s gay subcultures like that described in chapter 1 at HMAS Creswell were not common, although there were unsubstantiated reports of a ring of gay officers who allegedly produced and shared pornographic material at RAAF Base Edinburgh in the late 1970s and early 1980s.[126] Navy ships, however, remained sites where men who identified as straight would seek sexual pleasure with gay men. Bill Lord recalls Colin Edwards's dramatic discharge from HMAS *Melbourne* in 1981. Lord describes the ship's captain as an incredibly conservative man, and when he learned about Colin Edwards he did not hesitate to discharge him. As for several other men alleged to have had homosexual relations but who were believed to be straight—the captain had them all transferred to other ships.[127]

One topic where gay ex-servicemen's experiences diverged was social networks. Some either did not know any other LGB service members or were too fearful to befriend LGB Defence members. RAAF cook Kenny Spence, who died of AIDS-related conditions in 1992, said in an oral history interview before his passing:

> I kept very much to myself. As I got to know gay people up here in Darwin [1980–84], they'd say: 'Oh, you're so lucky; you're in the Force; you've got all these wonderful men at your feet.' But it wasn't worth it. There's enough guys outside the Air Force to keep you interested rather than be hassled by guys in the Force.[128]

Other ex-servicemen describe forging friendships with numerous LGB people across the three services, although these memories are more common among officers. Captain Tony McLeod explains that a gay mate from Duntroon 'would take me out to visit his friends, so he kick-started my gay social life. Whereas before it had been secretive, and now I was actually living in Sydney, going out on the town and meeting people and having friends who were gay.'[129] Of course, there were other ranks who also forged social networks with other gay servicemen. Colin Edwards recalls befriending several gay sailors at Sydney gay bars from 1978 to 1981.[130] WRAAC Corporal Chris Sinclair recalls her friend Dennis hosting fabulous parties and being a social hub for LGB service members in Canberra in the early 1980s. Unfortunately, however, one of those parties became his downfall when some military police showed up and subsequently began an investigation.[131]

Conclusion

As this chapter shows, the period 1974–92 was dangerous for LGB service members. New policies that were alleged to treat them with 'sympathy and discretion' codified procedures that authorised surveillance, interrogations and ultimately the discharge of hundreds of LGB service members throughout the ADF. The investigations in the Army especially targeted women, and the RAAF was disproportionately attempting to weed out all of its LGB members. The investigations devastated LGB members, who in some cases took months or even years to rebuild their self-esteem and find new career opportunities. Notwithstanding the ban, investigations and witch-hunts, LGB people continued to serve and found different ways to explore and express their sexuality.

One reason for the new policies and heightened vigilance was, as the 1974 policy review noted, the increasing visibility and shift towards societal tolerance of homosexuality. The gay and lesbian rights movement began in the 1970s and by the 1980s was attaining concrete gains such as law reform and some state anti-discrimination protections. Indeed, the service police seemed to crack down harder on LGB service not in spite of but because of the shifts in societal attitudes and a sheer determination to maintain the ADF's rigid gender and sexual norms. As chapter 4 explores, the very rationales for the ADF ban and its practical implications came under pressure from the early 1970s until it was finally overturned in November 1992.

Notes

1 'So you want to be a WRAAF', p. 5.
2 'Inquiry into WRAAF sackings', *Canberra Times,* 11 July 1973, p. 12; 'Transcript of segment of Mike Willesee's "A Current Affair"', CTC 7, 17 July 1973, available from the WRAAF's service record.
3 'Policy on Homosexuality in the Services', 11 May 1974, NAA A6721, 1985/18156 PART 1.
4 See chapters 6, 9 and 12 of Willett, *Living Out Loud.*
5 'Policy on Homosexuality in the Services: Report by the Principal Administrative Officers Committee (Personnel)', May 1974, NAA A6721, 1985/18156 PART 1.
6 'Policy Regarding Homosexuals in the Armed Services', 14 November 1974, NAA BP1028/2, 701/P1 PART 2. See also 'Policy Regarding Homosexuals in the Armed Services', 22 October 1974, NAA C1273, 705/P1; NAA BP899/6, 701/P1 PART 2.
7 'Homosexual Behaviour—Guidelines for Investigation', May 1974, NAA A6721, 1985/18156 PART 1; NAA C1273, 705/P1.
8 DI(G) PERS 15–3: 'Homosexual Behaviour in the Australian Defence Force', 4 November 1985.
9 'John', interview with Noah Riseman, 29 September 2015, Hobart.
10 Ex-RAAF policeman, letter to *OutRage*, 1991 (no date), available from ALGA.
11 'Pencil', interview.
12 Draft DI(A) PERS 96–1 Homosexual Behaviour, 5 November 1992, NAA A6721, 1985/18156 PART 4.
13 See Senate Committee on Foreign Affairs, Defence and Trade, 'Sexual Harassment in the Australian Defence Force', August 1994; Major Kathryn Quinn, Australian Army Psychology Corps, 'Sexual Harassment in the Australian Defence Force', December 1996.
14 'Terri', interview with Noah Riseman, 19 October 2015, Sydney.
15 Colonel P.J. Desmond, 'Brief for DCGS: Alleged homosexual activities', 4 April 1989, NAA A6721, 1985/18156 PART 1.
16 Yvonne Sillett, interview with Shirleene Robinson, 29 August 2016, Daylesford; SIB-M-75-88—Homosexual Activity—Y.M. Sillett, 1988, courtesy Yvonne Sillett; Riseman, Robinson and Willett, *Serving in Silence?*, pp. 148–51.
17 'Mark', interview with Noah Riseman, 4 March 2016, Sydney; Riseman, Robinson and Willett, *Serving in Silence?*, pp. 167–9.
18 Janet Carter, interview with Shirleene Robinson, 9 August 2016, Perth.
19 'Summary of Homosexual Offences', July 1982, NAA BP1028/2, 705/101/P1 PART 10.
20 Elaine Crump, interview.
21 Kyle Duffield, interview with Noah Riseman, 19 April 2017, Bicheno, Tas.

22 'Homosexual Behaviour—Guidelines for Investigation', May 1974, NAA A6721, 1985/18156 PART 1; NAA C1273, 705/P1.
23 Kate Carlisle, interview with Noah Riseman, 10 December 2017, Washington, DC.
24 Tony McLeod, interview with Noah Riseman, 20 April 2017, Bicheno, Tas.
25 Shane Duniam, interview with Noah Riseman, 29 May 2017, Lismore, NSW.
26 'In defence of the realm?', *OutRage*, March 1991, p. 17. See also Martyn Goddard, in 'Discriminating service', *Four Corners*, 3 August 1992, NAA C475, 1746037.
27 Danny Liversidge, interview with Noah Riseman, 22 January 2018, Melbourne.
28 'In defence of the realm?', p. 20.
29 Shane Duniam, interview. See also Kenny Spence, January–February 1992, in Dino Hodge, *Did You Meet Any Malagas? A Homosexual History of Australia's Tropical Capital* (Nightcliff, NT: Little Gem Publications, 1993), p. 203.
30 'Interview Report: LACW [redacted] Stwd, Base Squadron RAAF Point Cook', 3 December 1981, in NAA ST7486/1, 705/RIC/P1 Part 1; Shane Duniam, email to Noah Riseman, 8 May 2019.
31 Telex with subject heading 'Search of Suspects—Homosexual Investigations', 12 June 1986, NAA B5679, 701/P1 PART 3.
32 Ex-RAAF policeman, letter to *OutRage*. See also 'Discriminating Service', NAA C475, 1746037; B. Baker, 'ADF hate policy', *Sydney Star Observer*, 21 October 1999, p. 11.
33 Karl Bryant, interview with Noah Riseman, 17 February 2016, Melbourne.
34 Chris Sinclair, interview.
35 Richard Gration, interview with Noah Riseman, 13 December 2016, Sydney; Riseman, Robinson and Willett, *Serving in Silence?*, pp. 134–5.
36 Keith Drew, interview with Noah Riseman, 13 August 2016, Darwin.
37 Eunice Coogan, interview with Shirleene Robinson, 24 August 2016, Gold Coast.
38 Statement Concerning Conduct of Interview of Richard Gration, no date. This document is based on both the handwritten and typed 'Report on Investigation by RAAF', drafted a few days after the initial interview. Documents courtesy Richard Gration.
39 'Record of interview between Corporal R.W. Pickersgill A323494 RAAFPOLIN and LAC Kyle Austen Duffield A325255 Clerk Administration at RAAF Police Section Williamtown on 21 February 1989', courtesy Kyle Duffield.
40 'Record of interview conducted between FSGT J.M. Leahy A13359 RAAFPOLIN and SNRACDT X ['Ryan'] at Headquarters RAAF Base Point Cook on 1 February 1982', courtesy Richard Gration.
41 'Record of interview conducted between FSGT J.M. Leahy A13359 RAAFPOLIN and Cadet R.C. Gration A323545 Cadet Academy RAAF Base Point Cook on Thursday 4 February 1982', courtesy Richard Gration.
42 'Peter', interview.
43 Alix Blundell, interview with Noah Riseman, 26 August 2017, Brisbane.
44 Alix Blundell, interview.
45 Reparations Payments Assessor, Defence Abuse Response Taskforce, to Alexandra Blundell, 18 September 2014; Colonel Glenn Wells, to Alexandra Blundell, 7 April 2017. Documents courtesy Alix Blundell.
46 Lieutenant General Angus Campbell, Chief of Army, to Alix Blundell, 25 July 2015, courtesy Alix Blundell.
47 Explanation of the Commonwealth Superannuation Corporation Invalidity Benefits and classifications is available from fact sheet MB03 07/13, https://www.csc.gov.au/members/advice-and-resources/factsheets-and-publications/militarysuper, retrieved 15 June 2020.

48 Lyndal Green, interview.
49 Susie Struth, interview. See also Riseman, Robinson and Willett, *Serving in Silence?*, pp. 112–14.
50 Leon Fry, interview.
51 'Interview Report: LACW [name redacted] STWD, Base Squadron RAAF Point Cook', 3 December 1981, NAA ST7486/1, 705/RIC/P1 PART 1. The names in the document are redacted, but Shane Duniam confirmed that this report is indeed about him and his friends.
52 Simon Hammond, '5 sacked in camp probes', *Truth* (Melbourne), 26 December 1981, p. 2.
53 Sandy Ashton, interview.
54 Hammond, '5 sacked in camp probes', p. 2.
55 'Standard Explanatory Position Relating to DI(G) PERS 15–3: Homosexual Behaviour in the ADF', 27 June 1986, NAA A6721, 1985/18156 PART 1; NAA B5679, 701/P1 PART 3.
56 Ronni Spencer, interview.
57 Sandra McInerney, interview with Shirleene Robinson, 8 December 2015, Canberra.
58 Kate Carlisle, interview.
59 Stuart Martin, interview with Graham Willett, 24 January 2017, Melbourne. See also Keith Scott, 'RAAF sacks AIDS carrier', *Canberra Times*, 30 October 1987, p. 10.
60 'In defence of the realm?', p. 18.
61 Clifford Allen, 'Homosexuality: Its nature, causation, and treatment', in Charles Berg and Clifford Allen, *The Problem of Homosexuality* (New York: Citadel Press, 1958), pp. 14, 18; John Wolfenden, 'Report of the Committee on Homosexual Offences and Prostitution', ed. Scottish Home Department Home Office (London: Her Majesty's Stationery Office, 4 September 1957), pp. 11–12.
62 'Policy on Homosexuality in the Services', 11 May 1974, NAA A6721, 1985/18156 PART 1. See also 'Policy Regarding Homosexuals in the Armed Services', 22 October 1974, NAA C1273, 705/P1.
63 Defence Instruction (Army) PERS 96–1, 20 October 1982, NAA A6721, 1985/18156 PART 1.
64 Colonel A.R. Craig, DPsych, 19 January 1985, NAA A6721, 1985/18156 PART 1.
65 LtCol J.D. Duff, DPE-A (Directorate of Personnel Employment—Army), 2 January 1991, NAA A6721, 1985/18156 PART 2.
66 Lt Col P.G. Gregor, SO1 Coord, 29 January 1991, NAA A6721, 1985/18156 PART 2.
67 Ex-RAAF policeman, unpublished letter to *OutRage*, ALGA.
68 'Draft DI(G) Homosexuality in the Australian Defence Force', 24 April 1985, said: 'The ADF policy on homosexuality is that when a member admits to or is proven to be involved in a homosexual lifestyle, this includes bisexual behaviour, that member is to be subject to an administrative discharge.' The reference to bisexuality was gone in all drafts from May 1985 onwards. NAA A6721, 1985/18156 PART 1.
69 Darran Clark, interview with Noah Riseman, 25 August 2017, Brisbane.
70 Cable from 23 Fd Regt, to RAYWFF/DEFARM Canberra, 2 September 1982, NAA A4090, 537/6/19.
71 'Simon', interview with Noah Riseman, 9 September 2018, Melbourne.
72 Janet Carter, interview.
73 See Delia Quigley, interview; Susie Struth, interview; Riseman, Robinson and Willett, *Serving in Silence?*, p. 114.
74 Group Captain H.F. Freeman, 'Homosexual activities', 11 August 1981, NAA BP1028/2, 705/101/P1 PART 7.

75 Colin Edwards, interview with Noah Riseman, 29 August 2017, Townsville.
76 Colin Edwards, interview.
77 See Peter Jones, 'Towards self-reliance', in *The Royal Australian Navy: A History*, ed. David Stevens (South Melbourne: Oxford University Press, 2001), p. 231; Mark Corcoran, 'The luckiest refugees', ABC, 20 June 2012, https://www.abc.net.au/news/specials/the-luckiest-refugees/, retrieved 22 February 2019.
78 Colin Edwards, interview.
79 Kenton McKeown, interview with Noah Riseman, 18 December 2018, Sydney.
80 'Peter', interview.
81 'Bruce', interview with Noah Riseman, 12 December 2018, Cairns.
82 Danny Liversidge, interview.
83 Colin Edwards, interview.
84 Gen Ford, interview.
85 'Mark' interview; Riseman, Robinson and Willett, *Serving in Silence?*, pp. 169–71.
86 'Nathan', interview with Noah Riseman, 21 January 2016, Melbourne.
87 Richard Williams, interview with Noah Riseman, 29 November 2018, Griffith, NSW.
88 Major S.N. Aird, 22 October 1992, NAA A6721, 1985/18156 PART 5.
89 Jennifer Jefferies, interview.
90 'Pencil', interview; Lyndal Green, interview.
91 'Brett', interview with Noah Riseman, 28 April 2017, Melbourne.
92 Neil Murray, interview with Noah Riseman, 31 May 2017, Sydney.
93 Ray Griggs, interview with Noah Riseman, 24 August 2018, Canberra.
94 'Request statistics regarding investigations around homosexuality', 26 April 2018, Defence FOI 358/17/18.
95 This Royal Commission was set up in 2012 after years of pressure on government to investigate child sexual abuse at religious and other institutions that looked after children in out-of-home 'care'. The Royal Commission also examined the ADF over sexual abuse claims relating to underage cadets at HMAS Leeuwin, the Army Apprentice School, Balcombe, and more recently in the ADF cadets. The Royal Commission's final report is available from https://www.childabuseroyalcommission.gov.au/. The final report relating to the ADF case study is 'Report of Case Study No. 40: The response of the Australian Defence Force to allegations of child sexual abuse' (Sydney: Royal Commission into Institutional Responses to Child Sexual Abuse, August 2017), https://www.childabuseroyalcommission.gov.au/sites/default/files/case_study_40_-_findings_report_-_australian_defence_force.pdf, retrieved 22 February 2019.
96 Interestingly, this mirrors US data on gender and discharges for homosexuality. See Steve Estes, 'Ask and tell: Gay veterans, identity and oral history on a civil rights frontier', *Oral History Review* 32, no. 2 (2005): 41.
97 Australian Senate, Estimates Committee B, Report to the Senate, '1992–93 Senate Estimates Hearing Committee B—23 September 1992 Questions on Notice', pp. 144–5. See also documents from September–October 1992 contained within NAA A6721, 1985/18156 PART 4 and PART 5.
98 General Peter Gration, 'ADF Homosexuality Policy', 23 June 1992, NAA A6721, 1985/18156 PART 4.
99 Documents from September–October 1992 contained within NAA A6721, 1985/18156 PART 4 and PART 5. The internal document adds to twenty-four discharges, whereas the figure reported to Senate Estimates was twenty-two. For statistics on women's employment in the ADF, see Christine Reghenzani, 'Women in the ADF: Six decades of policy change (1950 to 2011)', Australian Parliamentary

Library Summer Scholar's Paper, 42, https://www.aph.gov.au/~/media/05%20About%20Parliament/54%20Parliamentary%20Depts/544%20Parliamentary%20Library/pubs/apf/scholarship/CReghenzani.pdf?la=en, retrieved 16 April 2019.

100 Deborah Cameron, 'Why Inga accuses the RAAF', *Sydney Morning Herald*, 19 November 1987, p. 4.

101 Documents from September–October 1992 contained within NAA A6721, 1985/18156 PART 4 and PART 5. In another inconsistency, one document indicates that the two discharged Army officers were male, but interviewee Captain Gen Ford was discharged during this period.

102 Colonel P.J. Desmond, 'Brief for DCGS: Alleged homosexual activities', 4 April 1989, NAA A6721, 1985/18156 PART 1.

103 Neil Murray, interview.

104 Michael Dinnison, interview with Noah Riseman, 30 August 2017, Townsville.

105 Brian Lawrie, interview with Shirleene Robinson, 20 January 2016, Sydney.

106 Sandy Ashton, interview.

107 'Bruce', interview.

108 'Bill', interview with Noah Riseman, 7 June 2017, Melbourne.

109 Brian Lawrie, interview.

110 See Neil Murray, interview; 'Kevin', interview with Noah Riseman, 11 December 2017, Washington, DC.

111 See New South Wales, 'Inquiry into gay and transgender hate crimes between 1970 and 2010', https://www.parliament.nsw.gov.au/committees/inquiries/Pages/inquiry-details.aspx?pk=2510#tab-members, retrieved 24 February 2019; State Coroner's Court of New South Wales, 'Inquest into the death of Scott Russell Johnson', 30 November 2017, http://www.coroners.justice.nsw.gov.au/Documents/Scott%20Johnson%20findings%20%20final.pdf, retrieved 24 February 2019; *Deep Water: The Real Story*, written by Amanda Blue and Jacob Hickey, directed by Amanda Blue, produced by Blackfella Films and Screen Australia, 2016, 87 min. Beat violence and other homophobic attacks were not confined to New South Wales. See for instance Alan Berman and Shirleene Robinson, *Speaking Out: Stopping Homophobic and Transphobic Abuse in Queensland* (Bowen Hills, Qld: Australian Academic Press, 2010).

112 'Kevin', interview.

113 Robert Reynolds, *What Happened to Gay Life?* (Sydney: NewSouth, 2007); Reynolds and Robinson, *Gay and Lesbian, Then and Now*, pp. 84–104; Wotherspoon, *Gay Sydney*, pp. 195–210.

114 Brian Lawrie, interview.

115 Craig Cahill, interview with Noah Riseman, 28 September 2015, Launceston.

116 Leon Fry, interview.

117 'Kevin', interview.

118 Delia Quigley, interview.

119 'Archibald', interview with Noah Riseman, 9 December 2018, Gold Coast.

120 Richard Gration, interview; Riseman, Robinson and Willett, *Serving in Silence?*, p. 130.

121 Craig Cahill, interview.

122 Tony McLeod, interview.

123 Colin Edwards, interview.

124 Keith Jeffers, interview with Noah Riseman, 16 November 2016, Canberra.

125 Tony McLeod, interview.

126 'RAAF porn claim to be investigated', *Sydney Morning Herald*, 15 May 1989, p. 12; 'Ex-RAAF staff alleged to belong to sex gang', *Canberra Times*, 15 May 1989, p. 5.

127 Bill Lord, interview with Noah Riseman, 25 September 2017, Canberra.
128 Kenny Spence, in Hodge, *Did You Meet Any Malagas?*, p. 203.
129 Tony McLeod, interview.
130 Colin Edwards, interview.
131 Chris Sinclair, interview.

Chapter 4

Challenging the ban

On 25 January 1971, the *Sydney Morning Herald* published a small article entitled 'Navy men discharged'. It read:

> Five sailors from HMAS *Swan* have been discharged from the Navy for homosexuality. The Minister for the Navy, Mr D.J. Killen, said tonight that the men had been discharged as 'unsuitable'. Mr Killen said that the swift discharge of the men was justified in the interests of the good name of the RAN, the protection of its men, from corruption, and concern for those who served in it.[1]

The next day, the co-founder of the gay and lesbian activist organisation Campaign Against Moral Persecution (CAMP) wrote a letter to Minister Killen condemning the sailors' dismissal. The letter attacked Killen's moralising attitude and noted: 'The homosexual is as innately qualified to serve his or her country as is the heterosexual. The Campaign Against Moral Persecution maintains that there is no evidence of the efficiency of the Navy suffering from homosexual activities within it.'[2] Killen sent a curt reply, dismissing CAMP's protest: 'I acknowledge your letter of 25th January and I have noted your views. I have nothing to add to what I have said publicly.'[3]

The letter from CAMP and the Defence Minister's response are emblematic of the long history of challenges to the ban on LGB military service. For CAMP and other LGB activists, the military represented one example of numerous institutions that denied access to LGB Australians. For more than thirty years, the military and government officials responded to criticism with their standard justifications for the ban, and the issue would usually disappear from the media just as quickly as it arose. Indeed, as the 1970s rolled into the 1980s, the military ban quietly disappeared from LGB activists' agenda as well. Ultimately, it would take the intervention of brave LGB Defence members and political allies to force the ban on to the public and political radar in the early 1990s.

This chapter examines the long fight to repeal the ban on LGB service, beginning with the low-key CAMP activism of the 1970s through to service members' efforts in the 1980s and finally to the legal challenges that placed the ban on the political agenda in 1992. It addresses the strategies deployed by activists and LGB service members, as well as how the ADF establishment adapted its justifications over three decades to defend the ban. Finally, the chapter explores the process that began when two lesbian servicewomen challenged their dismissals in the Human Rights and Equal Opportunity Commission in 1991, setting in motion the political and legal wrangling that ended with the lifting of the ban in November 1992.

Perhaps what is most remarkable about Australia's experience with the military ban, particularly when contrasted with the United States and the United Kingdom, is how little activism there was and how utterly *un*remarkable the decision to permit gay and lesbian military service became. Even so, overturning the ban required a host of unsung heroes who bravely challenged accepted policy and ingrained assumptions about military culture and LGB people's place in it.

CAMP activism

CAMP was not the only lesbian and gay rights organisation in 1970s Australia, but it was the only one that even sporadically discussed military service. Smaller organisations like the Daughters of Bilitis (Melbourne) and Homosexual Law Reform Association (Canberra) were focused on local or targeted issues, while the more radical Gay Liberation and Radicalesbians had no interest in military service. Yet even CAMP did not make the military ban a prominent target. Among the thirty-seven issues of the newsletter *CAMP Ink* published between 1970 and 1977, the topic only came up five times:

1. the February and March 1971 issues about the dismissal of the five RAN sailors described above
2. the 1973 article 'So you want to be a WRAAF' discussed in chapter 3
3. a 1973 first-hand account where 'Hadrian' described his service in the Australian Army
4. republication of RAN orders against homosexuality, which had been leaked to a CAMP member in 1973

5 reproduction of a 1975 letter sent from the Department of Defence to CAMP, explaining and justifying its ban on homosexuality.[4]

The minimal Australian activism seems remarkable when juxtaposed with what happened in the United States, from where the Australian gay and lesbian rights movement derived many of its strategies and discourse. The 1970s in the United States was a period of federal lawsuits from dismissed servicemen and women, most famously Leonard Matlovich, Vernon 'Copy' Berg and Miriam Ben-Shalom.[5] The Matlovich case alone received more press in Australia than the nation's own ban.[6] Especially since the 1954 Supreme Court case *Brown v. Board of Education*, the United States had a tradition of forcing civil rights action through the judicial system, and the military activism of the 1970s followed this tradition (albeit unsuccessfully). The Constitution of Australia does not lend itself to judicial activism, particularly with regard to civil rights, which means that organisations must lobby for change through the parliamentary system. The scale and concentration of the gay and lesbian communities also directed priorities differently in both countries. Activists in the United States began founding national gay rights organisations in the 1970s. In contrast, Australia's newly emerging associations remained state-based (albeit with some coordination across state lines) and prioritised securing state law reforms to decriminalise homosexual acts.

Still, CAMP members raised the military ban in relevant forums to prosecute its broader case for gay and lesbian equality. One such example was Western Australia's 1974 Honorary Royal Commission to Inquire into Matters Relating to Homosexuality. The state's Legislative Council established the Honorary Royal Commission as an extension of a joint select committee to examine a bill to decriminalise homosexual acts.[7] 'Hadrian', an anonymous CAMP WA member who had previously published a first-hand account of his Army experience in *CAMP Ink*, appeared before the Honorary Royal Commission as the witness 'C'. He submitted his *CAMP Ink* article and testified about his experience as a practising homosexual since the age of fourteen. He argued for the importance of a liberal society permitting private acts between consenting adults and that criminalised homosexuality left gay men vulnerable to blackmail and bashings. He also noted his many homosexual encounters in the Australian Army, summarising: 'You flatter homosexuals and homosexuality if you imagine that the practice is like a drug which turns straight people into addicts.'[8]

Almost seven weeks after Hadrian's testimony, Major General Cedric Maudsley Ingram Pearson, Chief of Personnel from the Department of Defence, also appeared before the Honorary Royal Commission to rebut Hadrian. The Defence submission emphasised that the armed forces did not permit homosexual members because they would adversely affect troop morale. Defence also published the key tenets of the new 1974 policy on investigating suspected homosexuals, outlined in chapter 3.[9] Although the Honorary Royal Commission recommended the decriminalisation of homosexuality in Western Australia, by the time it delivered its findings there was a new state government that did not support law reform. As for the military ban, the Honorary Royal Commission was the first time that the Defence establishment publicly espoused that it had a policy which was supposedly treating gays and lesbians 'sympathetically' and 'with discretion'.

Other attempts at lobbying for reform met with similar platitudes from the political establishment. W. Stankovic of Campus CAMP, University of Queensland, sent a letter to Prime Minister Gough Whitlam in June 1974 calling for the public service to end discrimination against homosexuals. Stankovic also pointed to 'Discriminatory Regulations & Rules—We request the repeal of these regulations, especially in relation to the Armed Forces and employment in security and classified areas'. Acting Prime Minister Jim Cairns responded by defending the ban on gay and lesbian service members: 'Life in the Services is generally felt to be an outdoor and vigorous one; one of conventionally normal and healthy endeavour. Acceptance and condonation of homosexual behaviour does not fit either the public or the Services view of the Armed Forces and would adversely affect both recruiting and re-engagement.' Cairns repeated the new Defence public line: 'Where cases do arise, the persons concerned are treated sympathetically and with discretion, and it is possible for them to leave the Services with an honourable discharge.'[10] Admittedly, at this early stage of the new tri-service policy, politicians did not know that the military would embark on witch-hunts and engage in intimidating investigation practices to target LGB service members. The political class considered the reformed policy to be sufficient and the matter closed.

The Royal Commission on Human Relationships

CAMP made one further intervention to challenge the military ban, which led to a symbolic victory. In September 1975, CAMP NSW sent a comprehensive, forty-page submission to the Royal Commission on Human Relationships. The Whitlam government had set up this inquiry to examine 'family, social, educational, legal and sexual aspects of male and female relationships' with the aims of informing public and social policy.[11] CAMP's submission contained the recommendation 'that the Commission, in representations to the Australian Government, support the right of homosexual defence personnel to live openly as homosexuals within the Defence

Services'.[12] To support its recommendation, CAMP reprinted the 'Policy Regarding Homosexuality in the Armed Forces' and provided a series of comments on the policy. For instance, CAMP argued that the supposed 'sympathy and discretion' was meaningless if it denied the right of gay and lesbian Defence members to live openly. CAMP also challenged the unproven premises underlying the policy, such as never explaining how gays and lesbians would affect morale, how their presence would erode public perceptions of the armed forces, the stereotype that gay men were unfit for the vigour of service and the supposedly rare occurrences of homosexuality in Defence.[13]

CAMP also pointed to the absurd way the policy attempted to differentiate between 'confirmed' homosexuals versus those 'unlikely to become involved in further incidents'. Queer theorists of the 1990s would distinguish between gay and lesbian as constructed identities versus those who perform homosexual acts but identify as heterosexual. Public health officials often refer to such individuals as 'men who have sex with men' or 'women who have sex with women'.[14] Yet, as CAMP was arguing, the arbitrary delineation showed 'total lack of awareness of the homosexual lifestyle' and, fundamentally, the policy represented 'a shallowness of understanding, a total lack of research, and a general insensitivity'.[15]

CAMP's submission had an effect. Contained within the Royal Commission's more than 500 recommendations was the following: 'The Defence Department should remove automatic discrimination against homosexuals in the services and judge their qualifications on the same criteria as would be applied to heterosexuals.'[16] The Royal Commission comprehensively documented most issues, including abortion, homosexual law reform, family planning, family violence and sexual assault. Yet there was no wider explanation for this recommendation beyond noting it as one of the numerous public institutions where gays and lesbians wrongfully faced employment discrimination.[17] The Royal Commission's recommendation was the first government inquiry to advocate lifting the ban on gay, lesbian and bisexual service. It was a minor recommendation in the wider context of the Royal Commission on Human Relationships, but one that did briefly garner press attention among the raft of other controversial proposals.[18] As Michelle Arrow has written, much of the contents of the Royal Commission's report became part of a political football, which the conservative Fraser government used to attack Whitlam and the ALP in the 1977 election.[19] CAMP advocated that citizens instead read the report, which had been 'cynically misrepresented by politicians and much of the press'.[20]

By 1980, well after the Royal Commission's full publication, the Fraser government was finally preparing its reply to the report's recommendations. Just as the recommendation on the LGB military ban was minor in the overall context of the Royal Commission, it received little specific attention from organisations responding to the recommendations. Only the South Australian Premier's response specifically endorsed the recommendation, which is not surprising given at that time South Australia was the only state to have decriminalised homosexual acts.[21] The Report of the Interdepartmental Working Group into the Royal Commission on Human Relationships briefly indicated: 'The Defence Force would seek exemption from the recommendations relating to homosexuality (R.VI.97, R.VI.98, R.VI.102) on the grounds that they have quite different requirements from those of the general community for maintenance of command relationships, discipline, morale and security.'[22] This claim that the ADF had different, inherent requirements that should exempt it from anti-discrimination protections would become more pronounced in the late 1980s and early 1990s.

A chart of 'Recommendations Considered to have Policy Implications', tabled in Cabinet on 14 May 1980, listed the four standard reasons given by the Department of Defence why they should have the discretion not to employ gays and lesbians: maintenance of discipline and morale, protecting minors employed in the services, public perceptions that homosexual acts did not belong in the ADF, and the possibility of blackmail or victimisation—an interesting rationale supposedly meant to protect gays and lesbians.[23] In the only public response to the Royal Commission recommendation, Minister for Home Affairs R.J. Ellicott endorsed Defence's position, stating:

> The Department of Defence takes the view that this is unacceptable on the grounds that the Defence Force has quite different requirements from those of the general community for the maintenance of command relationships, discipline, morale and security. Additionally, a high proportion of the Services comprises young persons for whom the Services have an obligation to act in loco parentis.[24]

The Royal Commission on Human Relationships—and the Fraser government's blatant dismissal of LGB military service—could have represented an opportunity to galvanise LGB activists and service members around the issue. Yet this did not happen. CAMP did not pursue the matter further, and the ban disappeared from the activist agenda, which focused on state law reform to decriminalise homosexual acts and draft anti-discrimination legislation. Only twice in the 1980s did gay rights activists write to politicians about the ban. The first occasion was in 1982 when Alexander 'Lex' Watson, former CAMP president and founding member of the NSW Gay Rights Lobby, wrote to gay-friendly ALP parliamentarians about the proposed Defence

Force Discipline Bill. He argued that the legislation did not sufficiently support LGB service, and instead that 'A liberalisation, working toward a total end to discrimination, is necessary in the defence forces'.[25]

The second was in November 1985 when the NSW Gay Rights Lobby wrote to Defence Minister Kim Beazley after the ADF introduced DI(G) PERS 15-3, updating the 1974 policy. Their letter debunked the four justifications for the ban and asserted: 'This is a most discriminatory Instruction which is out of keeping with present community attitudes and current trends towards anti-discrimination policies in the workforce.' The Navy Commodore who handled the letter described it to a colleague as 'Attached for your information and possible entertainment'. The draft response to the NSW Gay Rights Lobby again deployed the ADF's standard reasoning, asserting that homosexuals would adversely affect combat readiness.[26] The ban again disappeared from the radar of activists, who were more concerned with the AIDS epidemic and continuing state discrimination. Moreover, unlike the United States, there was not a stream of dismissed service personnel joining LGB rights organisations or pushing to challenge the ban.

GESA versus the RSL

LGB Defence members remained mostly silent throughout the 1970s and 1980s. In fact, the first issue of the new gay and lesbian publication *Campaign* in 1975 had an announcement calling for current or recently discharged gay, lesbian and bisexual service personnel to share their stories for an upcoming feature article.[27] This story never came to fruition, suggesting that the callout did not elicit sufficient responses. Throughout the 1980s the gay and lesbian press would sporadically publish short stories about military service and the treatment of suspected gays and lesbians. One letter from 'a reluctant soldier' exposed the blatant homophobia he had witnessed in Army basic training, indicating that if he were to come out, he 'would be reviled, ridiculed, bashed and abused by the men I share a room with'.[28] A gay ex-soldier replied three months later with the advice:

> If you want to stay in the army, I suggest you try for promotion. Then you will achieve the privacy of your own room or share with one other. You must admit that the store that you serve in gives you the biggest opportunity for window shopping there is! But until you can find a friend who is not connected with the services, you will feel lonely. Meanwhile, laugh at their sick humour and you will survive.[29]

The response did not address the ban as a structural issue to tackle, but instead saw it as a personal dilemma that the reluctant soldier had to confront. In a similar example, in 1982 *City Rhythm* briefly published a story about the surveillance and intimidation practices targeting 'John'. While condemning the ban, the article did not encourage LGB Defence members to challenge it. The article said:

> If you are in the forces, do keep yourself above suspicion. Do not keep any suspect articles such as porn, photos or letters, these could only lead to trouble for yourself and others. It is unfortunate that, for the time being, if you are gay and decide that the armed forces are your life, then you cannot be fully yourself.[30]

In early 1982, the Commonwealth Government was drafting the *Defence Force Discipline Act* to create a uniform military code for all three services and to bring ADF personnel under the jurisdiction of the Australian Capital Territory Criminal Code. The press reported that the legislation would also have a more liberal approach to homosexuals.[31] Within days the Department of Defence clarified that private homosexual acts off base would still be prohibited.[32] The short-lived proposal divided ex-service organisations, with RSL National President Sir William Keys stating, 'What you do in your private life, in a private area is your own affair—providing that it doesn't become the subject of public comment and doesn't affect your standard of work in the service.'[33] Outspoken Victorian RSL President Bruce Ruxton drew more attention when he stated, 'I am appalled, shocked and disappointed that this should happen.'[34]

Ruxton was known for decades as an outspoken conservative commentator on a raft of issues ranging from homosexuality to women's rights to Aboriginal land rights.[35] His high profile and vocal opposition to causes including gay rights have left many older LGBTI ex-service members still wary of joining the RSL. What is less known, though, is that within the upper echelons of the RSL leadership there were several closeted gay members. One national president had secret relationships with men,[36] as did Ruxton's immediate predecessor as Victorian RSL president, Colin Keon-Cohen. When Keon-Cohen died in 1982, his male partner of ten years was shocked to learn that he had not been included in the will. The gay magazine *OutRage* outed their relationship when the partner accepted a small settlement from Keon-Cohen's family in July 1984.[37] Closeted veterans like Keon-Cohen were not making public homophobic statements, but neither were they advocating for the rights of LGB active or returned service members. Instead, the public face of the RSL remained a bastion of homophobia until well into the 2000s. LGB ex-service members had to look elsewhere for support and recognition.

One night in March 1982, a group of five gay ex-airmen were discussing some of Bruce Ruxton's homophobic remarks. Max Campbell, who had served as a RAAF education assistant from 1961 to 1981, recalls: 'I just said fairly innocently, "Why don't we form some sort of group, association or something that we can get a little bit political in that we'll maybe do a wreath on Anzac Day or be prepared to be interviewed and that if someone asks?"'[38] Out of this statement the Gay Ex Services Association (GESA) was born. GESA placed announcements in the mainstream and gay press, with one announcement stating: 'It is intended to discuss the proposed aims of—a Social Group, Counselling, First Line Support, Back-up Support, Long Term Pressure Group.'[39] Ultimately only about eight to twelve people joined GESA (reports conflict), mostly gay men but also two lesbians. These young men and women had served in Vietnam or in Australia from the 1960s to the 1980s. The organisation mainly functioned as a social group, with monthly meet-ups in a pub.[40] GESA planned to lay a wreath at Melbourne's Shrine of Remembrance on Anzac Day 1982, with the ribbon accompanying their wreath saying: 'For all our brothers and sisters who died during the wars. Gay Ex Services Association.'[41]

Laying the wreath did not go smoothly. Five GESA men arrived at the Shrine just before the public wreath-laying at 12.30 pm. It was a staged event with a photographer from the gay monthly magazine *City Rhythm* present. The five men were noticeably gay because of their manner of dress: one in leather, others in jeans and one known as Teddy Bear Terry carrying a teddy bear in his back pocket. As the five men climbed the steps to the forecourt, Bruce Ruxton charged across and shouted: 'Stop those men!' Ruxton and the Shrine commissionaire refused to let the men enter and lay the wreath, even though they had previously received verbal undertakings from the RSL's deputy secretary that they could lay a wreath at the public time. The commissionaire summoned the police, who escorted the men away.[42]

From Ruxton's perspective, the Anzac legend could not encompass gays and lesbians. Ruxton even commented, 'I don't mind poofters in the march but they must march with their units. We didn't want them to lay a wreath because we didn't want to have anything to do with them. We certainly don't recognise them and they are just another start to the denigration of Anzac Day.'[43] Ruxton also deemed gays incompatible with digger masculinity when he remarked, 'You know as well as I do, they [gay soldiers] couldn't hide themselves. The men would get on to it straight away.'[44] Ruxton not only positioned GESA outside the imagined community of ex-servicemen and women but also grouped them with the Women Against Rape in War protesters as part of 'a concerted effort by anti-heritage groups to destroy the march'.[45] Women Against Rape in War had organised a small wreath-laying in Melbourne in 1979 and drew more than 600 protesters in Canberra in 1981. They would continue to protest Anzac Days in other cities throughout the 1980s.[46] The few historians who have written about the GESA incident also group them with Women Against Rape in War as part of a movement in the 1980s to challenge Anzac Day's glorification of war. Yet there is a clear distinction between these groups; GESA was challenging the *exclusivity* of the Anzac legend rather than challenging the day itself. GESA president Mike Jarmyn commented: 'We are not a political extremist group bent on the degradation of the Anzac Day tradition. We simply wish to publicly recognise the fact that gay people also gave their lives in war. We are not playing politics.'[47]

GESA continued its monthly meetings after the wreath-laying incident, and the following year it had to decide whether to attempt to lay another wreath. Four of the original five did not want to risk rejection again, but Teddy Bear Terry agreed to accompany Max Campbell to lay a private wreath. This time Campbell attracted the eye of Bruce Ruxton; he recalls:

> And he [a constable] started to take a bit of notice, so we just talked amicably for a little while longer. He went back and reported to Bruce. And, because I had established every right to be there and to—you know, everything—and oh, if looks could kill Bruce Ruxton would be a murderer. [Laughs] But I waited to the appointed time, I checked that no one was coming or going, and off I went with no problem whatsoever. I just walked in and I spent my moment and walked out, walked away.[48]

Campbell's wreath-laying did not attract the same media attention as 1982, partly because the Women Against Rape in War protests overshadowed it, but more so because it did not become a public incident.[49]

In 1984, Max Campbell laid the final GESA wreath. Again, Shrine commissionaires (although not Ruxton) turned Campbell away. A reporter was nearby and asked Campbell what was happening, and he calmly explained that he was being turned away for being gay. The journalist contacted Shrine staff, and later in the day rang Campbell to say he was welcome to return and lay the wreath.[50] The incident did not receive coverage in the gay press and had a brief mention in the *Sydney Morning Herald*, with the Shrine Chairman, Wing Commander Peter Isaacson, stating, 'The association [GESA] was allowed to lay a wreath last year and should have been allowed this year.'[51] GESA disbanded sometime after Anzac Day 1984 owing to lack of continuing interest among its members to continue the wreath-layings and monthly meet-ups. GESA never actively challenged the LGB military ban in its public statements or actions; rather, its symbolism aimed to draw attention to the fact that there had always been LGB people serving with honour, thus challenging the mythology underpinning the ADF's defence of the ban.

Service members challenge the ban

For those LGB men and women who were serving in the 1970s and 1980s, there was little recourse within the ADF system to contest the ban. That said, there were rare circumstances when connected or savvy service members challenged the procedures that service police were deploying to surveil, investigate, intimidate and discharge them. The most dramatic case of this occurred at the RAAF Academy in Point Cook, Melbourne, in early 1982. A third-year cadet was caught after he tried to climb into the bed of another cadet with whom he had previously had sexual relations. The sleeping cadet was no longer interested in a relationship and reacted angrily, prompting the perpetrator to jump out of a ground floor window and run across the courtyard naked. When RAAF Police investigated, they compelled the accused to name seven other cadets with whom he had sexual relations, including Richard Gration.[52]

Richard had joined the RAAF Academy in 1981, and he had connections that most other Defence members did not: his father and uncle were high-ranking Army and RAAF officers respectively. From 1984 to 1987 his father Peter would serve as Chief of the General Staff—the present-day Chief of Army—and from 1987 to 1993 he was Chief of the Defence Force (CDF). Richard's uncle Barry was Chief of Air Force from October 1992 to 1994. RAAF Police summoned Richard for an interview just as he had returned from an eight-day training exercise. After a few hours of pressure, Richard described his three sexual encounters with the other cadet. Only after Richard confessed did the formal interview commence, highlighting one of many inappropriate RAAF Police procedures. Richard, the original cadet and three others subsequently received notice that they were to be discharged from the RAAF.

The details about inappropriate police conduct emerged only because, unlike almost all other cases of accused homosexuals, Richard fought the recommendation for discharge. Peter Gration arranged for Richard to meet Brigadier M.J. Ewing, Director of Army Legal Services. Ewing surprised both Peter and Richard by suggesting that they fight the dismissal on the grounds that Richard's sexual encounters were isolated experiences and that he was not a confirmed homosexual. Ewing prepared a legal brief advising that Richard's entire interview was potentially inadmissible because the police did not follow proper rules of gathering evidence.[53] Richard, of course, did self-identify as gay, but he did not disclose this to Ewing or his father.[54]

Ewing contacted Defence Minister James Killen, advising him that RAAF Police had mistreated the cadets. Killen accepted Ewing's advice but worried that since everyone at the RAAF Academy knew about Richard and the three other accused cadets' homosexual encounters, they would not be comfortable serving alongside them. One evening the four accused (not the original perpetrator) summoned all Academy cadets to a meeting. As Richard recalls:

> We explained the situation of 'Look, the Minister considers that it would be problematic that none of you would accept us ... if you are happy for us to stay on in these circumstances we'd be very grateful if you'd write a statutory declaration saying that you're aware that something is being investigated but you're more than happy for us to stay on.' And so we got 120 stat decs; we got the whole Academy did it [*sic*].

The family of one of the other accused cadets gave the statutory declarations to their member of parliament, who forwarded them to the Defence Minister. Then, as Richard puts it, 'The shit absolutely hit the proverbial fan.'[55] Minister Killen set up a Court of Inquiry into the Conduct of RAAF Police at the Air Force Academy.

The Court of Inquiry sat in April 1982. The legal officer representing Richard and the other cadets used civilian and military precedents to argue that the interview conditions were unreasonable and constituted collecting evidence under duress. The ADF's legal officers argued that there was no misconduct in the investigation and that even the defendants were not challenging the findings of fact.[56] The Court of Inquiry's final report simultaneously vindicated and criticised the RAAF Police. It found that the RAAF Police's findings of facts were mostly accurate and 'that the interviews were generally conducted properly and in accordance with the relevant rules and the required procedures contained in RAAF publications'. But the Court of Inquiry also determined that there had been small procedural misjudgements around such matters as cautioning the cadets about how their interviews would be used, the timing and rushed nature of the interviews, the specificity and accuracy of the interview transcripts, and unclear directions about the presence of a commissioned officer. Taken collectively, these minor procedural actions disadvantaged the cadets. The Court of Inquiry concluded that the standards of evidence for an administrative process must adhere to the same rules as a disciplinary process.[57]

The Court of Inquiry's findings were forwarded to the Attorney-General's Department and to the Defence Minister. The Crown Solicitor advised the Chief of Air Staff against discharging the cadets, arguing that any discharges could be challenged in an embarrassing Federal Court case.[58] Richard and the other cadets were not discharged and instead only received written warnings not to engage in any homosexual conduct again.[59] This Court of Inquiry is the first known challenge to service police for their excessive practices investigating cases of homosexuality. The intervention of such high-profile personalities was clearly a factor, but the outcome remains telling. The ruling underlined that the intimidating nature of police practices and methods of gathering evidence in cases of suspected homosexuality were at best inappropriate and at worst unlawful. Yet, as the numerous examples provided in chapter 3 reveal, the Court of Inquiry did not lead to widespread change in police practices.

Other service members who did not have powerful allies searched for other means to appeal their own cases or to challenge the ban. The Commonwealth Government established the Commonwealth Ombudsman in 1977, and in 1984 legislative amendments transferred the Defence Force Ombudsman (DFO), set up in the Department of Defence in 1975, into the Commonwealth Ombudsman's office.[60] The first gay services-related complaint to the Commonwealth Ombudsman was in 1980. In 1979, there was a break-in at a post office adjacent to a RAN establishment. When matching the tampered mail with the recipients, one Navy official found a letter that identified one sailor as gay and reported him to a superior officer. At the time the Commonwealth Ombudsman did not have jurisdiction to investigate the Department of Defence. The Ombudsman's annual report for 1980–81 concluded: 'In this case the complainant wanted to be reinstated in the Navy but because I was unable to investigate the Navy's actions after it became aware of his homosexuality it was not open to me to suggest that he be reinstated.'[61]

Even when the DFO did have the jurisdiction to investigate complaints after 1984, it could only focus on whether the policy and procedures had been followed. After receiving several complaints from gay and lesbian Defence members, the DFO wrote to the CDF in December 1988 challenging the logic of the ban.[62] In his response to the DFO, General Peter Gration wheeled out the standard justifications for the ban and added:

> The rank structure in the ADF also means subordinates may be dissuaded from or reluctant to initiate complaints against a superior for fear of the possible consequences. The posting and reinforcement requirements of the ADF mean that practising homosexuals, who have the potential to create these difficulties, should be excluded from the entire Force, not just from the combat elements.[63]

It is intriguing that the CDF argued that the ban almost protected gays and lesbians from bullying. Incidentally, the CDF's argument could apply to any intimidating behaviour that discouraged subordinates to challenge higher ranks. Notwithstanding these illogical claims from the CDF, the DFO was satisfied with the response, concluding: 'In general I have few problems with the ADF's stated policy, although I believe much of the statement is conjectural rather than evidential.'[64]

The DFO correspondence with the CDF, like the RAAF Academy Court of Inquiry, did not challenge the ban itself so much as the unjust procedures used to implement it. As chapter 3 relates, the vast majority of LGB service members investigated never had the opportunity to challenge their cases. Later, there was a new jurisdiction where LGB service members could lodge complaints: the Human Rights and Equal Opportunity Commission (HREOC). In 1990 that statutory body's terms of reference were widened to include sexual orientation, giving them the authority to investigate government and workplace discrimination complaints.

The case of Lucy Kardas, a straight woman investigated for being a lesbian in 1991, reveals that complaints to the HREOC could produce concrete outcomes for specific complainants but at great cost. In early 1992 Kardas discovered that the RAAF Police had secretly been investigating her for homosexuality. She is convinced that the false allegation came from superior officers as revenge for when she reported colleagues for sexual harassment and fraud. In October 1991, an anonymous complainant reported coming 'upon a female officer "tongue kissing" another female' and identified the officer as Lucy Kardas. The RAAF Police secretly interviewed other service members who were present at the officer's mess on the night in question. Based on those interviews, the investigator determined that 'the evidence appears to substantiate the allegation of homosexuality, and as the incidents I have investigated involve such behaviour in a public context, I believe if you decide to take administrative action then you have no choice, in accordance with RAAF policy, but to recommend her discharge'.[65] Kardas's commanding officer did not accept the report's findings, particularly because Kardas and another witness had never been interviewed, so he ordered a new RAAF Police investigation. This second investigation concluded that there was no evidence to substantiate the allegations that Kardas was a lesbian.[66]

Kardas was never questioned in either of these investigations; she only found out about the second one when a civilian friend who had been interviewed contacted her. Her commanding officer admitted that the investigation had been going on for months and advised Kardas that she had been cleared. Still, the RAAF refused to release any further information to her. Kardas contacted a civilian lawyer and then filed a complaint with the HREOC, alleging that the secret investigations violated her human rights and that she had been subjected to sexual harassment. While the ADF could not legally prevent its members from going to the HREOC, according to Kardas the organisation strongly discouraged it, including requiring members to notify commanding officers if they lodged any claims. What ensued was a three-year battle with the RAAF, which put immense strain on Kardas's mental health and resulted in a suicide attempt. She was subjected to bullying, ostracism, poor performance appraisals and denial of professional development opportunities.

As a result of the HREOC complaint, five of the RAAF investigators were censured, and Kardas also received notification that the original complainant admitted to having fabricated the whole story. The HREOC process finally concluded in 1995 when the RAAF agreed to a settlement of an undisclosed amount and a medical discharge for Kardas.[67] Air Vice Marshal D.N. Rogers, the then Deputy Chief of the Air Staff, wrote to Kardas: 'The fact that you were cleared of any wrong doing whatsoever was not formally conveyed to you for some time and that is a matter of concern and regret to me. I offer an unreserved apology both for the unsatisfactory administration of the matter on the part of the RAAF and the suffering and distress that you have experienced.'[68]

All of the above cases reveal how the internal, hierarchical structure of the ADF left little room for service members to challenge policy—and even then only in exceptional circumstances. Lucy Kardas's case also shows how straight allies could be caught up in the web of investigations and secrecy that shrouded the ban on LGB service. Even heterosexual ADF members who opposed the ban had no opportunities to raise or challenge it, as decisions were made from the top and the ADF had a culture that discouraged questioning decision-making processes. Ultimately, as all of the above cases highlight, it would take intervention from outside the ADF—be it through ministers, the DFO or the HREOC—to force the ADF at the very least to defend its policies and, in exceptional circumstances, to reverse individual cases. Eventually, it would be also outside forces that finally brought down the LGB ban.

ADF members' attitudes towards homosexuality

Whether Defence members would welcome the presence of LGB people was always speculative, with both pro- and anti-ban advocates using anecdotal evidence to fuel their arguments. In 1991 the ADF conducted some consultation forums and a survey to gauge members' attitudes towards homosexuality and the LGB ban. The Department of Defence Personnel Division noted markedly different attitudes in the three services. The Army, focused on combat, still opposed LGB service as threatening morale. The RAN expressed a similar attitude for ship deployments, but thought shore-based roles might not inherently require heterosexual personnel. The RAAF advised that there was no inherent job requirement to exclude homosexuals because the organisation was less tight-knit; instead, 'Air Force's concern was centred on the prejudice which would be directed against homosexuals.'[69]

A more systematic appraisal of soldiers' attitudes came from the Ready Reserve Soldier Attitude and Opinion Survey, conducted with a random sample of other ranks from the Army. The survey used a Likert scale, where every question had five possible answers: strongly disagree, disagree, undecided, agree, strongly agree. In May 1992, for the first time, the survey included the question: 'I would not mind serving alongside homosexuals so long as their sexual behaviour doesn't interfere at work.' With a 55 per cent response rate, 71 per cent of respondents answered strongly disagree and 9 per cent answered disagree.[70] As chapter 5 shows, subsequent surveys even after the ban was lifted continued to show high disapproval rates on the question about serving alongside homosexuals.

These reports consistently suggested that Defence members did not want to serve alongside LGB people, and oral histories both reinforce and challenge this data. Several interview participants remember suspected or known lesbians and gay men being targets of abuse and harassment in the 1970s and 1980s. Ronni Spencer endured taunts of 'lesbian' while serving at HMAS Platypus in the mid-1980s, while Matt Cone recalls a sailor from basic training in 1992 who was ostracised because of his camp demeanour.[71] 'Mark' recalls one suspected gay soldier being thrown down the stairs and two who were bashed when others discovered their sexual relationship (c. 1987–88).[72] Two interview participants who served in very different times and places remember the practice of 'blanket bashing'. 'John M.', who served in the Army in the 1970s, explains that soldiers 'would throw a blanket over them [suspected gays] and people would get stuck into them so they would never know who was hitting them'.[73] A decade later, in 1987, sailor Kenton McKeown was falsely accused of attempting to fondle another sailor's genitals. He recalls that one night:

> I woke to already having a blanket over my face, a hand on my neck and having the shit beaten out of me and punched in the balls, the cock, my stomach and my chest. I remember someone saying either on that beating or another beating, 'Don't hit his fucken arms. Someone hold his legs, someone hold his arms, so we don't bruise them. The bruising can't be visible.' And then as quick as it happened, the curtain was closed and then the blanket was pulled away after the curtain was closed. That happened about

> five or six times with words like, 'Fucken faggot, we are going to fucken kill you. You don't belong here. You know what happens to faggots.'[74]

Both John M. and Kenton McKeown suffered even more severe abuse than just the blanket bashings. The back of John M.'s head was split open in one bashing, and he was sexually abused, which led to him contracting venereal disease; for years he turned to alcohol to cope with the trauma he endured in the Army.[75] For McKeown, the abuse that began with blanket bashing escalated until one afternoon a group of about ten sailors threw him overboard while HMAS *Derwent* was docked at HMAS Stirling, south of Perth. The fall was eleven metres with an impact speed of 106 kilometres per hour. Fearing for his life, McKeown told commanding officers it was a just a prank that went wrong, and for decades he kept the secret of what really happened.[76] These examples are not isolated incidents. Several of the recent inquiries into the history of physical and sexual abuse in the ADF found that perpetrators often targeted 'other males on the basis of sexuality or perceived sexuality'.[77]

The culture of homophobia in the ADF had reverberations in civilian Australia as well. 'Geoff' was a closeted gay man who served in the Australian Army from 1984 to 1991. He says that while he was based at Holsworthy in western Sydney, colleagues would regularly go into the inner city and harass gay men. He recalls even discreetly warning some effeminate men to leave a bar because he feared that his Army friends might bash them.[78] 'Jason' also remembers that one soldier whom he jerked off at Kapooka 'used to say [that] him and his mates used to go poofter bashing in Sydney. [They'd attack] guys who go off doing beats or whatever.'[79] In 1988, the media reported on a spate of gay-bashings in Townsville allegedly perpetrated by Army and RAAF members. The gay community even had a nickname for the aggressive, homophobic abusers: AJs, for Army jerks. In one case, a gay man living with AIDS was beaten so badly that his spleen was ruptured and he had to be flown to Brisbane for emergency treatment. Notwithstanding numerous allegations from Townsville's gay community (and Aboriginal community as well) over hate crimes, ADF representatives denied that their members were perpetrating homophobic or racist violence.[80]

Although the above examples and surveys suggest a strong homophobic culture in the ADF, other oral histories suggest that this was not uniform throughout the services. Chapter 3 presents some cases where ally service members either warned or protected their LGB friends from investigations. The above-mentioned story about the RAAF Academy incident showed that 120 cadets were willing to sign statutory declarations supporting four suspected gay men. Other anecdotes reveal colleagues who discovered their friends' sexuality and were supportive. Leigh Easterbrook, who served in the Army from 1982 to 1986, remembers a colleague who saw her holding another woman's hand in a news clip about the 1986 Sydney Gay and Lesbian Mardi Gras. She remembers the colleague saying, '"Oh, I saw you on the news the other day, Corporal, last night." And I went, "Oh, yeah", and I thought, "Oh." And he goes, "No, it's all right", you know, and he was really all right.' On another occasion, Easterbrook blurted out to a colleague that she was gay and burst into tears. His response was to pat her on the back, close the door and say: '"It's all right. It's nothing to be embarrassed about." I said, "I'm not embarrassed about it." He goes, "It's all right. My sister is gay; that's all right."'[81] Several close airmen friends knew that Keith Jeffers was gay when he served at RAAF Base Edinburgh in Adelaide in the early 1980s. His boyfriend even used to go out with his workmates, and he shared a house with a man known to be gay.[82]

These and other examples suggest that although the ADF had a broadly homophobic culture, it was not uniform, and individual friendships could challenge prejudice. This is what sociologists refer to as the contact hypothesis: that meeting members of groups considered 'other' can challenge stereotypes and discrimination and change people's attitudes.[83] Indeed, as more LGBTI people have gradually come out since the 1970s, Australian society has undergone remarkable attitudinal shifts, most obviously manifested through the 61.6 per cent 'yes' vote for marriage equality in the 2017 postal survey.

For the ADF, which had a longstanding culture of mateship and valued initiative and competency,[84] strong leadership could ensure that LGB personnel would be welcomed into the organisation. This was not forthcoming from the service chiefs, though. Minutes from a March 1992 Chiefs of Staff Committee meeting 'noted the disquiet expressed by the Service Chiefs of Staff in the light of feedback from their Services ... about the prospect of changes to the current policy on homosexuality in the Australian Defence Force'.[85] In July 1992, Warrant Officer First Class Arthur Francis, Regimental Sergeant Major of the Army Land Command, told the Defence Minister at a public forum that the whole Army wanted the ban preserved and that soldiers would leave if LGB people were allowed to serve.[86] It was therefore clear by 1992 that the ADF hierarchy was unwilling to revoke the ban on LGB military service; only political intervention would force their hand.

Lifting the LGB ban

In 1991, shortly after sexual orientation was added to the HREOC's terms of reference, the commission received complaints from two dismissed lesbian servicewomen: one from the Navy and one from the Army. Although the

HREOC had no legal grounds to compel these women's reinstatement, after eighteen months of conciliation the ADF did agree to restore their employment.[87]

Meanwhile, the HREOC investigated the ban on lesbian and gay service. There were no federal anti-discrimination statutes that prohibited employment discrimination on the basis of sexual orientation. Instead, the HREOC turned to international law, arguing that the ban contravened Australia's obligations under International Labour Organization (ILO) conventions and, more significantly, the International Covenant on Civil and Political Rights (ICCPR). Over the next fourteen months the HREOC and ADF negotiated a new policy on unacceptable sexual behaviour in the ADF, which the HREOC envisaged would be neutral as to sexuality.[88]

In late February 1992 the press reported that the ADF was considering lifting the ban. The government responded that discussions were still under way with the service chiefs. On 18 June 1992, Defence Minister Robert Ray announced in the Senate that the ban on gay and lesbian service would remain.[89] Within a few days there were already reports of dissent from other Cabinet ministers, most notably Attorney-General Michael Duffy. The HREOC and Human Rights Division within the Attorney-General's Department persuaded Duffy that the ban contravened Australia's obligations under the ICCPR, to which Duffy was firmly committed.[90] The HREOC highlighted three particular sections: the right to privacy (article 17), the right to access to public service (article 25) and the right to equality before the law and equal protection of the law (article 26).[91] Within days of Ray's announcement of 18 June, Australian Democrats Senator Janet Powell introduced amendments to the *Defence Act 1903* that would have overturned the ban. She withdrew the amendments because they did not have sufficient support, but later in the year she threatened to introduce similar amendments again and to force a vote.[92]

It is intriguing that reform advocates invoked international law to argue against the LGB ban considering that only a small number of nations, mostly in Western Europe, permitted LGB people to serve.[93] Throughout 1992 both sides of the debate frequently looked to overseas examples. Opponents of the ban talked about the successful integration of LGB troops in some Western European militaries; supporters of the ban pointed to the Anglosphere—the United States, New Zealand, United Kingdom and Canada—which all banned homosexuals from their militaries (although even those nations were debating their bans). There were questions about whether the United States in particular would permit joint exercises with Australia if the ban were lifted. The service chiefs indicated that precedents in NATO suggested that this would not be a problem.[94]

Duffy himself acknowledges that while there was a case to be made about international law, really it was an excuse to bring the issue out of the exclusive purview of the Defence Minister. For Duffy and others seeking to repeal the ban, it was a moral issue about ending discrimination against LGB people.[95] Now confronted with conflicting views from multiple ministers and their departments, the government faced clear divisions.

To defuse the issue, Prime Minister Paul Keating established a Caucus Joint Working Group on Homosexual Policy in the Australian Defence Force. Chaired by Senator Terry Aulich and with five other ALP parliamentarians from across factions, the Caucus Joint Working Group was an attempt to build a consensus within the ALP to resolve the matter. The Caucus Joint Working Group carefully deliberated all the standard arguments defending the ban—troop morale, protection of minors, health concerns and threats of blackmail—and most importantly considered the views of the service chiefs. The Caucus Joint Working Group also received submissions and welcomed testimony from gay rights advocates.[96] Rodney Croome recollects preparing a 114-page submission on behalf of a coalition of state gay and lesbian rights organisations, drawing material from his international networks as well. Croome says that one unintended positive consequence of the work on the submission was the formation of the Australian Council for Lesbian and Gay Rights as a national LGB rights organisation.[97]

The debates over the LGB military ban attracted some attention in the mainstream press, and the ABC aired a special studio debate as an episode of *Four Corners*.[98] The ban was certainly of interest to Australia's LGB community, as evidenced by the regular coverage in the LGB press. Letters to the editor show that LGB Australians overwhelmingly wanted the ban lifted, seeing it as another important step on the march to equal rights.[99] LGB activists staged a small day of action in July 1992 in Sydney, Melbourne, Perth, Brisbane, Canberra and Hobart outside ADF recruiting centres and ALP offices. In both Perth and Hobart, activists also laid wreaths at the cities' main war memorials. The Perth wreath commemorated LGB people who died in war and the Holocaust, while the Hobart wreath read: 'In memory of all the lesbians and gay men who have died in the war against homophobia and in all wars.' These were small demonstrations, though, with the Hobart one attracting only about a dozen people.[100] Some politicians such as Aulich took notice of LGB activists' positions, but the government's decision-making was mostly working independently of any outside influences.

On 18 September 1992 Aulich handed down the Report of the Caucus Joint Working Group. In a 4–2 split, the report advocated repealing the ban on gay and lesbian service as well as the implementation of training and education programs to facilitate a smooth transition.[101] Describing the split, Aulich recollects that the two dissenters simply did not think that ADF members were ready for the ban to be lifted. He also describes the entire Caucus Joint Working Group as 'a fair process in place no matter where we were coming from in terms of our initial views'.[102] Although the report carried significant clout for its methodical examination of the issue, because the committee did not attain a consensus, Ray and Duffy had to prepare two Cabinet submissions arguing their cases for and against the ban respectively.

At last the ban on LGB service went to Cabinet on 23 November. Duffy stressed the international law case, and he also mentioned the imminent repeal of Canada's ban. Ray retorted that while the ban was admittedly discriminatory, the ADF lawfully discriminated on multiple grounds including age, fitness and ability. The ban was necessary to maintain *esprit de corps*, and international covenants did not apply equally to defence forces.[103] After Duffy and Ray presented their cases, Keating at last laid his position on the table: he did not support any delay tactics and agreed the ban should go. Keating recollected in 2015: 'It made sense to me to actually support the right of gay people to be part of the armed services, to be who they were and to say who they were. So I came down on that side of the debate.'[104] When the debate concluded, a majority of Cabinet supported the decision to repeal the ban.

On 23 November 1992, Keating issued a statement announcing the end of the ban: 'This decision reflects broad support in the Australian community for the removal of employment discrimination of any kind, including discrimination on grounds of sexual preference. The decision brings ADF policy into line with the tolerant attitudes of Australians generally.'[105] The Minister for Defence Science and Personnel, Gordon Bilney, asserted: 'These international obligations, which are supported by all parties in Parliament, do not represent some alien rules forced on us; rather, they are the embodiment of the principles Australians believe in, and a symbol of the enlightened country we believe Australia to be.'[106]

The LGB press celebrated the Cabinet decision, and the *Age* hailed it as ending one of the remaining 'bastions of discrimination' and bringing the ADF into 'the 20th Century, a little later than most Australians'.[107] The Opposition disagreed with the decision and announced that their policy would be to follow the advice of the service chiefs, even if that meant reinstating the ban. As Aulich indicates, for the sake of policy consistency the service chiefs said that they would not seek a reversal of Cabinet's decision.[108] The RSL condemned the decision, but high-profile World War II prisoner-of-war Weary Dunlop supported the government, stating: 'There have always been homosexuals in the services. Don't let us delude ourselves … It is a mistake if you start labelling people too hard.' Duffy credits Dunlop's pronouncement as silencing the RSL and other critics of the reform.[109] As

Duffy and others in Cabinet foreshadowed, there was no measurable electoral backlash over the repeal of the ban. By 1994, the Coalition, too, accepted that LGB military service was part of the ADF landscape.[110]

The Keating government's November 1992 decision coincided with similar moves around the world. On 27 October, the Federal Court of Canada ruled that the Canadian Defence Force's ban on homosexuals violated the nation's Charter of Rights and Freedoms.[111] Gays and lesbians were already allowed and even required to serve in the Israeli Defence Force under National Service, but they were denied many security clearances. An announcement in October 1992 explicitly welcomed homosexuals and removed security restrictions.[112] The Canadian and Israeli reforms were independent of Australia, as were the American debates that began with Bill Clinton's election in November 1992 and culminated in the implementation of the 'don't ask, don't tell' policy in February 1994.

One nation where Australia's reform did have an impact, though, was New Zealand. There, a proposed Human Rights Bill would ban discrimination based on sexual orientation, which would have the extra effect of overturning the New Zealand Defence Force (NZDF) LGB ban. As recently as July 1992 New Zealand's Defence Minister had advocated for the bill to exempt the military, but by December 1992 the NZDF changed its position and would no longer seek an exemption. The Human Rights Bill passed in August 1993 and marked the end of the NZDF's LGB ban.[113]

As Bilney observed, these rapidly evolving policies globally brought Australia 'into line with most of our Western allies whom we fought alongside in the Gulf War last year'.[114] Australia was neither ahead of nor behind the international community, but rather was with the pack.

Conclusion

On 24 November 1992, the CDF sent out a message, which stated:

> Australians, regardless of their sexual orientation, will now be able to serve their country in the ADF without restriction ... The central element of the new policy is that sexual relations and activities are primarily and predominantly a private matter for each individual, and that the ADF has no concern with the sexual activities of its members provided they are not unlawful and are not contrary or inconsistent with the inherent requirements of the ADF.[115]

As this chapter shows, this final order to permit LGB people to serve openly came after more than two decades of courageous advocacy from LGB activists, service members, veterans and allies. Whether testifying before royal commissions, writing letters to politicians, sending anonymous letters to the press, demonstrating in public or confronting their chains of command, these brave men and women all challenged discriminatory policies and entrenched homophobic attitudes within the ADF. When the ban eventually was put on the political agenda, Attorney-General Michael Duffy championed the push for its repeal, overcoming divisions within his own party.

Interestingly, though, most interview participants who served when the ban was lifted have no memory of that historical moment, or have only vague memories of seeing an announcement or some discussion in the mess hall. 'Simon', who was serving on HMAS *Moresby* when the ban was lifted, even goes as far as to say: 'I don't remember it actually being much of a thing. I reckon it was more of a thing that Sydney won the Olympics, because "the winner is Syderney [*sic*]", that was the big thing.'[116]

Reflecting the views of many other ex-service members, former RAAF member Craig Cahill summarises: 'It was a non-event, in my opinion. I felt relieved because I don't have to deny it if anyone asked me, but there was still a lot of homophobia. Homophobia doesn't vanish overnight.'[117] As chapter 5 shows, this was most certainly the case for many LGB Defence members. Although they could no longer be dismissed because of their sexuality, being openly gay, lesbian or bisexual was a still risky business for the next decade. The 1990s and early 2000s proved to be an era when courageous LGB service members and allies would need to challenge deeply entrenched cultures and policies that continued to discriminate.

Notes

1 'Navy men discharged', *Sydney Morning Herald*, 25 January 1971, p. 3. See also 'Sailors out', *Canberra Times*, 26 January 1971, p. 3.

2 John Ware, to the Hon. D.J. Killen, 25 January 1971, reproduced in 'For Queen and country', *CAMP Ink*, February 1971, p. 12.

3 D.J. Killen to J. Ware, 5 February 1971, reproduced in 'Killen Replies', *CAMP Ink*, March 1971, p, 11.

4 'So you want to be a WRAAF?', pp. 4–5; Hadrian, 'Homosexuality in the forces', p. 4; 'Royal Australian Navy orders', *CAMP Ink* 3, no. 7 (1973), pp. 4–6; 'Letter received from the Department of Defence', *CAMP Ink* 4, no. 5–6 (1975), pp. 11–12.

5 Shilts, *Conduct Unbecoming*, esp. pp. 205–372; Eric Marcus (ed.), *Making History: The Struggle for Gay and Lesbian Equal Rights, 1945–1990, an Oral History* (New York: HarperCollins, 1992), pp. 279–92; Estes, *Ask and Tell*, pp. 185–97; Hippler, *Matlovich, the Good Soldier*; Gibson, *Get Off My Ship*.

6 See for example Robert Pullan, 'Homo hero in battle with the system', *Brisbane Telegraph*, 27 May 1975, p. 16; Lesley Oelsner, 'Homosexual's challenge to air force', *Canberra Times*, 11 June 1975, p. 18; 'Gay airman wins new hearing', *Herald* (Melbourne), 9 December 1978, p. 12.

7 Western Australia, *Report of the Honorary Royal Commission Appointed to Inquire Into and Report Upon Matters Relating to Homosexuality*, 18 September 1974.

8 'C', testimony to Western Australia, *Honorary Royal Commission Appointed to Inquire Into and Report Upon Matters Relating to Homosexuality*, transcript vol. 1, 2 May 1974, p. 41.
9 Cedric Mausley, testimony to Western Australia, *Honorary Royal Commission Appointed to Inquire Into and Report Upon Matters Relating to Homosexuality*, transcript vol. 3, 25 June 1974, AU WA S1989—cons5527 3, pp. 589–91; 'Letter received from the Department of Defence', *CAMP Ink* 4, no. 5–6 (1975) pp. 11–12.
10 NAA A451, 1974/6782.
11 Arrow, *The Seventies*, pp. 141–68. On the relationship between homosexuality and the Royal Commission on Human Relationships, see Arrow, '"These are just a few examples of our daily oppressions": Speaking and listening to homosexuality in Australia's Royal Commission on Human Relationships, 1974–1977', *Journal of the History of Sexuality* 27, no. 2 (2018): 234–63.
12 CAMP NSW, *Homosexuals and Human Relationships: Submission to the Royal Commission on Human Relationships* (Sydney: CAMP NSW, September 1975), pp. 19, 27.
13 Ibid., p. 31.
14 Rebecca M. Young and Ilan H. Meyer, 'The trouble with "MSM" and "WSW": Erasure of the sexual-minority person in public health discourse', *American Journal of Public Health* 95, no. 7 (2005): 1144–9.
15 CAMP NSW, *Homosexuals and Human Relationships*, p. 31.
16 Commonwealth of Australia, *Royal Commission on Human Relationships Final Report*, vol. 1 (Canberra: Australian Government Publishing Service, 1977), p. 124; ibid., vol. 5, p. 138.
17 Commonwealth of Australia, *Royal Commission on Human Relationships Final Report*, vol. 1, p. 87.
18 'Govt attacks plans for big change to sex laws', *Sydney Morning Herald*, 1 December 1977, pp. 1–2.
19 Arrow, *The Seventies*, pp. 187–93.
20 'What the report said!', *CAMP Ink*, no. 28, January 1978: 18.
21 NAA A1209, 1977/503 PART 7, p. 87.
22 NAA A12909, 3954, pp. 37–8.
23 NAA A12909, 3989, pp. 29–30; NAA A1209, 1977/503 PART 7, p. 149; NAA A10756, LC1781 PART 2, pp. 58–9; NAA A12909, 3954, pp. 109–10.
24 NAA A10756, LC1781 PART 2, p. 56; NAA A12909, 3989, p. 26.
25 Lex Watson, 'Defence Force Discipline Bill 1982', 25 May 1982, Alexander 'Lex' Watson collection, State Library of NSW (courtesy Robert French).
26 NAA A6721, 1985/18156 PART 1.
27 'Are you in HM Forces?', *Campaign*, September 1975, p. 5.
28 A reluctant soldier, 'Hate in uniform', *Campaign*, December 1978, p. 4.
29 R.H., Tarraville, Vic, 'Try for promotion', *Campaign*, March 1979, p. 4.
30 'No gays beyond this point!', *City Rhythm* (Melbourne), February 1982, p. 23.
31 'Department "will not accept" homosexual behaviour', *Canberra Times*, 21 January 1982, p. 11; 'New code to govern homosexuals in the forces', *Sydney Morning Herald*, 20 January 1982.
32 'Defence Dept fiddles law', *Klick!* (Melbourne), March 1982, p. 5; 'Don't they know? Defence Department confusion over status of gays', *Star* (Sydney), 12 February 1982, pp. 2, 7.
33 'New forces code turns blind eye to off-base sex', *Campaign*, February 1982, p. 6.
34 'Ruxton attacks service reforms', *Gay Community News* (Sydney), March 1982, p. 8.
35 Anne Blair, *Ruxton: A Biography* (Crows Nest, NSW: Allen & Unwin, 2004).

36 Kerry O'Sullivan, interview with Noah Riseman, 28 February 2018, Sydney. O'Sullivan had a relationship with a former RSL president. Although the former president is now deceased, we have not named him here because he has never been outed.
37 Danny Vadasz, 'RSL President disinherits gay lover', *OutRage*, July 1984, p. 5.
38 Max Campbell, interview with Noah Riseman, 13 March 2015, Melbourne.
39 'GESA', *City Rhythm*, March 1982, p. 3; 'Gay ex-services association', *Klick!*, April 1982, p. 7.
40 From June 1982 to October–November 1985, the 'City Rhythm's guide to gay Melbourne … and beyond' listed GESA's monthly meet-ups at the pub. See *City Rhythm*.
41 'RSL boss slights ex-servicemen', *Campaign*, June 1982, p. 6.
42 'RSL head provokes shrine skirmish', *City Rhythm*, June 1982, pp. 13–15; Max Campbell, interview.
43 Mary-Louise O'Callaghan, 'Ban on homosexuals' wreath bid', *Age*, 26 April 1982, p. 3.
44 Alison Thorne and Noel Maloney, 'Anzac Day: Women, gays can't lay wreaths', *Gay Community News*, May 1982, p. 5.
45 'RSL president sparks Anzac row', *Campaign*, May 1982, p. 5.
46 Christina Twomey, 'Trauma and the reinvigoration of Anzac: An argument', *History Australia* 10, no. 3 (2013): 98–106; Mark McKenna, 'Anzac Day: How did it become Australia's national day?', in Lake and Reynolds, *What's Wrong with Anzac?*, pp. 117–18; Holbrook, *Anzac*, pp. 120–1; Bruce Scates, *A Place to Remember: A History of the Shrine of Remembrance* (Cambridge: Cambridge University Press, 2009), pp. 244–7.
47 'Ruxton's ruckus at Shrine', *Klick!*, May 1982, p. 33. See also 'RSL head provokes Shrine skirmish', pp. 14–15.
48 Max Campbell, interview.
49 'Gays lay wreath', *Outrage*, May 1983 p. 8; 'Showdown at the Shrine: Violent arrests mark Anzac Day protests in Sydney and Melbourne', *Outrage*, May 1983 pp. 8–9; 'Anzac Day', *Campaign*, May 1983, pp. 10–11.
50 Max Campbell, interview.
51 Greg Roberts, 'Old soldiers parade, but it's a day of protest for some', *Sydney Morning Herald*, 26 April 1984, p. 3.
52 Richard Gration, interview; 'Proceedings of a Court of Inquiry into allegations made against RAAF Police by RAAF Academy cadets', 29 April 1982, NAA B4586, 704/1572/P1 PART 1; Record of interview conducted with cadet (name withheld to protect privacy) on 1 February 1982; RAAF Police statement by cadet, p. 3. February 1982. Documents courtesy Richard Gration. The story is also available in Riseman, Robinson and Willett, *Serving in Silence?*, pp. 120–9.
53 Legal opinion by Brigadier M.J. Ewing, DALS, to Major-General Peter Gration, 15 March 1982, courtesy Richard Gration.
54 Richard Gration, interview.
55 Ibid..
56 'Transcript of proceedings at RAAF Fairbairn on Monday, 19 April 1982, at 1502 hours', p. 1122, courtesy Richard Gration.
57 'Proceedings of a Court of Inquiry into allegations made against RAAF Police by RAAF Academy cadets', 29 April 1982, NAA B4586, 704/1572/P1 PART 1.
58 Author handwriting unclear, 'Note of Action', 30 July 1982, NAA B4586, 704/1572/P1 PART 1.
59 'Departmental Formal Warning, ACDT R.C. Gration A323545', 9 July 1982, courtesy Richard Gration.

60 Commonwealth Ombudsman, *Defence Force Ombudsman: Twenty-five Years of Service* (Canberra: Commonwealth Ombudsman, 2009), pp. 1–3, https://www.ombudsman.gov.au/__data/assets/pdf_file/0018/36207/defence_force_ombudsman-twenty-five_years_of_service.pdf, retrieved 5 December 2019.
61 Commonwealth Ombudsman, *Commonwealth Ombudsman Fourth Annual Report 1980–81* (Canberra: Australian Government Publishing Service, 1981), p. 31.
62 NAA A6721, 1985/18156 PART 1.
63 Commonwealth Ombudsman, *Commonwealth Ombudsman and Defence Force Ombudsman Annual Reports 1988–89* (Canberra: Australian Government Publishing Service, 1989), p. 191.
64 Ibid., p. 117.
65 'Investigation into Allegations of Homosexual Behaviour Involving FTTLT L.K. Kardas L320676', 21 November 1991, courtesy Lucy Kardas; Lucy Kardas, interview with Noah Riseman, 19 September 2017, Geelong.
66 Major General S.N. Gower, Assistant Chief of the Defence Force—Personnel, to ACT Human Rights Office, 11 March 1994, courtesy Lucy Kardas.
67 Joint Standing Committee on Foreign Affairs, Defence and Trade (Defence Subcommittee), 'Inquiry into Military Justice Procedures in the Australian Defence Force', submission no. 46: Lucy Kardas, 6 May 1998; Joint Standing Committee on Foreign Affairs, Defence and Trade (Defence Subcommittee), *Official Hansard Report*, 18 May 1998, pp. 85–97; Lucy Kardas, interview.
68 Air Vice Marshal D.N. Rogers, Deputy Chief of the Air Staff, to Lucy Kardas, 9 January 1995, courtesy Lucy Kardas. Kardas's memory poses interpretive questions about the significance of the archival trail. See Riseman, 'The interview–document nexus', pp. 27–31.
69 'Review of Discriminatory Policies and Practices: Stage 1 Discrimination on Grounds of Sex and Sexual Preference', October 1991, NAA A6721, 1985/18156 PART 3.
70 Lt-Col R.C. Furry, 'The Ready Reserve Soldier Attitude and Opinion Survey', p. 7 September 1992, NAA 1985/18156 PART 4. See also Captains J.S. Salter and S.E. Hodson, 'Soldier Attitude and Opinion Survey: Topical items—1992', Canberra: 1st Psychological Research Unit, Australian Army, 1995, pp. 5–6.
71 Ronni Spencer, interview; Matt Cone, interview with Noah Riseman, 27 May 2017, Port Macquarie; Riseman, Robinson and Willett, *Serving in Silence?*, p. 207.
72 'Mark' interview; Riseman, Robinson and Willett, *Serving in Silence?*, pp. 164–5.
73 'John M.', interview with Graham Willett, 18 January 2017, Wallan, Vic.
74 Kenton McKeown, interview.
75 John M., interview.
76 Kenton McKeown, interview.
77 Defence Abuse Response Taskforce, *Final Report* (Canberra: Commonwealth of Australia, March 2016), p. 30. See also Rumble, McKean and Pearce, *DLA Piper Review*, pp. 1, 19.
78 'Geoff', interview with Noah Riseman, 26 September 2016, Sydney.
79 'Jason', interview with Noah Riseman, 28 October 2016, Melbourne.
80 Greg Roberts, 'RAAF, Army deny bashings claims', *Sydney Morning Herald*, 30 August 1988, p. 4. See also Ed Bailey, interview with Noah Riseman, 18 March 2017, Canberra.
81 Leigh Easterbrook, interview with Shirleene Robinson, 9 February 2017, Sydney.
82 Keith Jeffers, interview.
83 P.J. Henry and Curtis D. Hardin, 'The contact hypothesis revisited', *Psychological Science* 17, no. 10 (2006): 862–8.

84 Lieutenant Colonel M.L. Phelps, 'The Australian Army's culture: From institutional warrior to pragmatic professional', *Australian Defence Force Journal* 123 (1997): 41.
85 Department of Defence, Chiefs of Staff Committee, minutes of meetings, 11 and 23 March 1992, Defence FOI 258/15/16.
86 'National day of action over military ban', *Melbourne Star Observer*, 17 July 1992.
87 NAA A6721, 1985/18156 PART 2 and PART 3.
88 Australian Human Rights and Equal Opportunity Commission, *Report of the Human Rights and Equal Opportunity Commission on Australian Defence Force Policy on Homosexuality* (Sydney: Craft Printing Industries, 1992), pp. 1–20.
89 Senator the Hon. Robert Ray, Commonwealth Parliamentary Debates (hereafter CPD), Senate, 18 June 1992, p. 3952.
90 Michael Duffy, interview with Noah Riseman, 1 May 2014, Melbourne; Margo Kingston, 'Defence gay ban prompts review by key ministers', *Age*, 22 June 1992, p. 4.
91 Australian Human Rights and Equal Opportunity Commission, *Report of the Human Rights and Equal Opportunity Commission on Australian Defence Force Policy on Homosexuality*, pp. 14–19.
92 CPD, Senate, 23 June 1992, pp. 4336–42; CPD, Senate, 7 October 1992, p. 1326.
93 Depending on the source, the number of nations varies between eight and fifteen. Some sources count countries that did not have an explicit ban, whereas others only include countries that specifically indicated they permitted homosexuals to join. See Stanley E. Harris, 'Military policies regarding homosexual behavior', *Journal of Homosexuality* 21, no. 4 (1991): 67–74; Hugh Smith, *Homosexuality and the Australian Defence Force: Individual Rights vs Organizational Realities*, Australian Defence Studies Centre Working Paper (Campbell, ACT: University College, University of New South Wales, Australian Defence Force Academy, 1992), pp. 21–3; David R. Segal, Paul A. Gade and Edgar M. Johnson, 'Homosexuals in Western armed forces', *Society* 31, no. 1 (1993), pp. 37–42; Peter Grogan and Rodney Croome, 'A submission to the ALP Caucus Committee on the Australian Defence Force Policy on Homosexuality', 1992.
94 *Report of the Caucus Joint Working Group on Homosexual Policy in the Australian Defence Force (ADF)*, September 1992, pp. 5–6; Terry Aulich, interview with Noah Riseman, 14 May 2014, Melbourne.
95 Michael Duffy, interview.
96 Terry Aulich, interview.
97 Rodney Croome, telephone interview with Noah Riseman, 14 December 2018; Grogan and Croome, 'A submission to the ALP Caucus Committee on the Australian Defence Force Policy on Homosexuality'.
98 'Discriminating service', *Four Corners*.
99 See for instance 'Ray's ban', *Capital Q* (Sydney), 17 July 1992; several letters published in *Sydney Star Observer*, 10 July 1992, and *Melbourne Star Observer*, 24 July 1992.
100 'Wreathes laid for gay and lesbian service personnel' and 'Gays protest outside Hobart recruitment office', *Pink Thylacines* (Tasmania), September 1992; 'Australia wide protest on ADF', *Westside Observer* (Perth), August 1992; 'GLAD ADF demo', *Melbourne Star Observer*, 24 July 1992; 'National action over military ban', *Capital Q*, 17 July 1992; 'Action on defence ban', *Sydney Star Observer*, 24 July 1992; Rodney Croome, interview.
101 *Report of the Caucus Joint Working Group*, p. 7. The education program was in line with recommendations from Gary Brown, 'Homosexuality and the Australian Defence Force: The issues', Background Paper No. 16, 1992 (Canberra:

Parliamentary Research Service, 14 August 1992), pp. 9–10.

102 Terry Aulich, interview.

103 NAA A14217, 868; NAA A14217, 880; NAA A11117, KCA535 Part 1; Michael Duffy, interview; Neal Blewett, *A Cabinet Diary: A Personal Record of the First Keating Government* (Kent Town, SA: Wakefield Press, 1999), pp. 267–70.

104 Kerry O'Brien, *Keating* (Sydney: Allen & Unwin, 2015), p. 622.

105 Paul Keating, 'Statement by the Prime Minister, the Hon. P.J. Keating, MP, Australian Defence Force policy on homosexuality', 23 November 1992.

106 Gordon Bilney, 'Statement by Minister for Defence Science and Personnel', 26 November 1992, ALGA.

107 'The military is now marching in step', *Age*, 25 November 1992, p. 13.

108 Alexander Downer, 'Media release: The defence force and homosexuals', 23 November 1992; Tim Fisher, 'Media release: Homosexuals in the defence force', 23 November 1992; Terry Aulich, interview.

109 Michael Millett, Alicia Larriera and Richard Macey, 'Now public service gays seek rights', *Sydney Morning Herald*, 25 November, 1992, p. 1; Michael Duffy, interview.

110 Alexander Downer, interviewed on *Sunday*, ABC, 3 July 1994.

111 'ADF ban: Australia isolated', *Melbourne Star Observer*, 13 November 1992, p. 3; 'US and Canada act to overturn ban', *Sydney Star Observer*, 11 December 1992, p. 13.

112 'Israel army recognises gays', *Capital Q*, 23 October 1992, p. 7; 'Israel army recognises gays', *Melbourne Star Observer*, 30 October 1992, p. 7.

113 Warren Cooper, New Zealand Minister for Defence, 'Memorandum for Cabinet. Human Rights Commission Amendment Bill', NAA A9737, 1992/13992 PART 1; cablegram from AUSTCOM Wellington to HQADF Canberra, 9 December 1992, NAA A1209, 1992/3382.

114 Bilney, 'Statement by Minister for Defence Science and Personnel', ALGA.

115 Telegram from CDF Canberra to various Army command units, 24 November 1992, NAA A6721, 1985/18156 PART 5.

116 'Simon', interview.

117 Craig Cahill, interview.

Chapter 5

An era of openness? 1993–2005

The Australian Government ended the ban on lesbian, gay and bisexual service in the ADF on 23 November 1992, but the lived experience of this change was uneven. While LGB servicemen and women were officially allowed to disclose their sexuality and remain in the services, their treatment within the ADF still ranged anywhere from complete acceptance through to being subjected to endemic bullying. Defence members still had to consider the decision to 'come out' and live openly LGB lives carefully.

Generally speaking, women and men who had established careers in non-combat roles were more likely to be received favourably. A range of policies continued to disadvantage LGB members, including broadly interpreted definitions of 'unacceptable sexual behaviour', opposition to diversity education programs, policies in relation to HIV/AIDS and the lack of recognition afforded to same-sex couples. However, the period beginning in 1993 was one of gradual change, culminating in reforms that included the recognition of same-sex relationships in late 2005. These changes were largely due to the efforts of courageous individuals and groups such as G-Force and DEFGLIS, who continued to challenge discrimination within the ADF after the ban was lifted. Transgender and gender-diverse Australians were still banned from serving openly in the ADF in this period, and their experiences are investigated in chapter 6.

Service life after the lifting of the ban

The removal of the ban meant that LGB servicemen and women could no longer be dismissed from the ADF because of their sexuality. It did not address or remove embedded prejudices that existed both within and outside the ADF. There is evidence that intolerance was still high. The annual Army Soldier Attitude and Opinion Survey asked a question about homosexuality from 1992 to 1995. Using a Likert scale, the survey asked respondents whether they agreed with the statement: 'I would not mind serving alongside homosexuals so long as their sexual behaviour doesn't interfere at work.' Consistently an overwhelming majority responded 'strongly disagree' or 'disagree', with the results compiled in table 5.

Table 5: Army Soldier Attitude and Opinion Survey results for statement 'I would not mind serving alongside homosexuals so long as their sexual behaviour doesn't interfere at work', 1992–95

Group surveyed	Strongly agree (%)	Agree (%)	Neither/ undecided (%)	Disagree (%)	Strongly disagree (%)
1992: Soldiers (other ranks)	2	11	7	9	71
1993–94: Soldiers (other ranks)	4	19	11	12	53
1995: Soldiers (other ranks)	4	23	17	11	45
1993: Officers	8	34	11	15	31
1994: Officers	6	42	11	15	25
1995: Officers	10	36	14	15	24

Three important points are evident in this data: (1) officers were consistently more accepting of LGB soldiers than other ranks; (2) the trend after lifting the LGB ban was generally towards support for LGB service; but (3) still, within three years of the ban's lifting, a majority (56 per cent) of other ranks and more than a third (39 per cent) of Army officers did not approve of LGB service. Gender breakdowns in some of the later surveys found that women were more likely than men to respond favourably to the statement.[1] Although attitudes would continue to shift slowly, as late as 2000 the national president of the RSL, Major-General Peter Phillips, said that 'there is no place for practicing [*sic*] homosexuals in the ADF' and asserted that 'overt homosexuals' would pose a 'clear threat' to 'teamwork and morale'.[2]

The homophobic attitudes expressed after the removal of the ban drew on longstanding stereotypes about masculinity and military service that have been explored in previous chapters. As sociologist Ben Wadham explains, 'The bonds of men in these arms corps and training environments are galvanised through othering and homosociality. These young men are rapaciously heterosexual and homophobic.'[3] Although Wadham does not suggest that all men who serve in the ADF are homophobic, he argues that military subjectivities have traditionally encouraged an idealised version of 'hegemonic masculinity, divorced from nature, emotion, woman and the savage; the captain of one's fate and the master of one's soul'.[4] Wadham's hypothesis would also explain why women, also marginalised by the ADF's masculine nature, would be more accepting of LGB people.

Ed Bailey, who served in the RAAF between 1990 and 1994, maintains that stereotypes about masculinity had powerful purchase within the ADF and that this might have prevented some people from disclosing their sexuality. The fear of being identified as homosexual and being stereotyped as 'not masculine' and as 'feminine' would have implications for service personnel. He asserted:

> I think it's that thing of obviously in the Defence Force you have to be a certain kind of type, like in the sense of male, masculinity and all that kind of stuff. So I think how you act and how you look, and how you perform—it was important to have that masculinity, and I think if you were anything apart from that general masculine look or act or anything else like that, you were considered to be something else. And I think a lot of people didn't want to be seen as that; they still wanted to put on that charade of, 'I'm still a man. I'm still a male', and there was always that thing of still pretending to be something that they weren't.[5]

David Wells had particularly strong ideas about the nature of masculinity, which kept him deep in the closet during his service from 1986 to 2000. Wells worked very hard to make everyone think he was straight, and he is still uncomfortable with some of the flamboyant stereotypes associated with gay life:

> So it's not that I have a problem with those people, it's just that—for some reason that's what the media think everyone is like and I'm not like that. I'm just a normal person. The only difference is I just have sex with guys, not girls … I don't really do anything different. You watch the footy, you drive fast cars and, you know, I just don't pick up the chicks, that's all. That's the only difference.[6]

While stereotypes and related discrimination were factors many service personnel had to consider before coming out, there were other reasons that LGB service personnel were cautious about revealing their sexuality. Service personnel had spent considerable energy in preceding years concealing their sexuality for fear of being forcibly discharged. Many LGB service personnel were therefore understandably cautious to react to this change in the immediate aftermath of this reform. 'Jason', who served in the Army from 1989 to 1998, explains, 'I thought it [being gay] would impede my career. That was just some perception. Not that anybody had ever said anything, but RSMs [regimental sergeant majors] are RSMs—they're so old school. And so then when I was in the Sergeants' Mess, it's like "You're one of the boys now, so don't do anything to fuck it up."'[7] Ed Bailey also recalls:

> I think people were still very much guarded and still very much in the closet. I don't think it's one of those openly gay things where you suddenly see people change in their mannerisms and openly discuss their sexualities in front of other people. I think it was just still that type of secret type of, 'I'm not going to tell anybody. This is taboo. I can't tell anybody that I'm …', and I think it's all about that identity.[8]

'Pencil', who served in the Army for twenty-five years before retiring in 1996, had been subjected to two investigations in her military career before the ban was lifted, and the treatment she was subjected to took a toll. She says when the ban was lifted in 1992, 'I don't really remember that actually happening, so it didn't make any influence on my life.'[9] Indeed, a surprisingly large number of interview participants do not even remember the ban being lifted. Sailor Kate Carlisle continued to keep her relationships secret, and remembers that if people asked about them she would reply: '"It's really none of your business", or if there was a friend who I … felt that I could be more open with, you know, I'd say. But I never really did. No, I kept that to myself.'[10] Pencil later reflected on the freedom she felt when she left the ADF in 1996: 'Well, I mean the day that I got out of the Army—that was just like this big tonne weight that was lifted off my shoulders.'[11] For service personnel who had hidden their sexuality for decades because of the costs of exposure, the concept of suddenly being able to be 'out' in the ADF must have appeared surreal and potentially risky.

Those interview participants who remember the ban being lifted tended to be officers, and they vividly describe the enormous sense of release they felt after the removal of the ban. 'Kevin' remembers, 'It was a great relief, actually, and I do remember thinking, "Thank God I don't have to worry about it."'[12] RAAF officer 'Bill' noted a mix of relief but also believes that service police were still targeting LGB people until they were finally ordered explicitly to stop. Bill did not come out right away, but he remembers taking small steps like bringing his partner to the wedding of an Air Force colleague.[13]

Every interview participant who was serving at that key moment in November 1992 feels that the homophobia in the ADF went unchallenged. Kevin states, 'It didn't change a lot, because people's views were the same; people were still going to call you names. And a lot of people weren't happy with the fact they lifted the ban, you know, and I'm sure some people would want it to come back.'[14] Bill recalls one fellow RAAF officer saying that he would wear a black armband and made some bizarre statements: '"We don't want these lemons in the military." "What do you mean by lemons?" And he was referring to lesbians, because apparently—and I said, "Why is that a lemon?" He said, "Well, because they're not good for anything." Appalling, appalling mindset, but that was a reasonably common thing then.'[15] In 1995, an anonymous lesbian officer who had served fourteen years in the Navy reported similar observations to the *Melbourne Star Observer*, asserting that 'even though the Government says you're allowed to be here, it doesn't mean that the people you work with can accept that or like it'.[16]

First-hand accounts from LGB service personnel also reveal that prejudice was embedded across the ADF and manifested in various ways. The story of Matt Cone shows just how difficult it could be to serve as an openly gay man in the years immediately after the ban was lifted. Cone joined the Navy in July 1992, months before the ban on LGB service would be lifted. After he finished recruit school and his steward training, he was posted to HMAS Watson in Sydney. Although he could no longer be discharged on the grounds of his sexuality, he soon learned to live a double life and to conceal his sexuality. He worked alongside 'some of the hardest, toughest sailors' who made it clear 'there should be two things that shouldn't be at sea: poofs and women'. He witnessed sailors participating in acts of homophobia. Once, they threw a beer bottle out of a car window as they drove down Oxford Street, then the epicentre of Sydney's gay social scene. Another time, they drove past a gay nightclub, the Albury, shouting: 'Faggot! You're dead cunts.'[17]

Cone was later the victim of several homophobic incidents, including being accused of 'unacceptable sexual behaviour' and 'prejudicial behaviour', which he linked to being identified as a gay man in the Navy. In 1996 Cone fronted a Defence magistrate's hearing over false allegations that he had groped a heterosexual sailor when, according to Cone, that sailor had come on to him. The Defence magistrate dismissed the case, but the entire saga had a detrimental effect on Cone's career and mental health as he was subjected to bullying and even death threats. In 1997 another Defence member drugged and raped Cone, and when he tried to report it, the divisional officer accused Cone of being a troublemaker. Cone had, by then, been completely traumatised by the scale of abuse he had experienced. The divisional officer put discharge papers in front of him, and Cone signed them. His treatment in the

Navy has had a continuous effect on his mental health, resulting in PTSD, anxiety and depression, and has affected his relationships and his employment opportunities.[18]

For more senior, heterosexual male officers, the lifting of the ban was viewed in retrospect as a success largely because it was a non-event. Upon his retirement in 1993, CDF General Peter Gration said, 'I won't comment on whether it was a good decision or not, but [there has been] very little or even no impact.'[19] A former Chief of Army, Lieutenant General David Morrison, does not recall encountering any challenges for openly gay men among his command roles in the 1990s. In fact, he fondly recalls a private from the assault pioneers who came out to his entire infantry platoon of about 700 men. Morrison describes the assault pioneers as some of the most macho men in the Army, and word spread that the private 'is one of theirs and anything that happens to him will be taken strongly by them'. Morrison also met personally with the private to assure him of his support. As far as Morrison knows, that soldier never encountered any homophobia and eventually climbed through the ranks to become a warrant officer.[20]

In 2000, Commodore R.W. Gates, a heterosexual senior warfare officer who had served in the Navy for twenty-nine years, gave an interview where he described the impact of the ban being lifted. His account is a more positive one than Matt Cone's. He said, 'I must admit, after it happened, it's been an absolute non-event. We've had some major cases of people declaring.' Commodore Gates related a memorable case:

> One of our executive officers of a destroyer, the second-in-command … declared. And, I'll be frank, it created a bit of a stir. We're talking about a mid-rank lieutenant commander in an absolute critical position on board a major warship—one heartbeat from command … That person under the new policy was certainly not removed from the ship, and in fact completed his full posting.

Commodore Gates described how the lieutenant commander approached the process of coming out in quite a measured way. He first discussed his decision and motivation with the ship's captain, outlining his desire to be truthful and act with integrity. Gates reported that his captain and other officers were generally supportive and respectful of his position as second-in-command. Furthermore, Gates noted the lieutenant commander had continued to serve successfully in the Navy and had been promoted to full commander.[21]

While Gates and Morrison were able to provide positive examples of Defence members coming out in the 1990s, these accounts are presented from positions of straight privilege. Other LGB veterans observed that homophobia was prevalent within the ADF of the 1990s. Vince Chong began his long RAAF career as a closeted ADFA cadet in 1997. He recalls one instructor describing homosexuality as 'legal but not compulsory' and witnessing other instructors suggest that homosexuals would be encouraged to discharge from the Army.[22] Colin Edwards, an ex-sailor who has been active in Townsville's gay community since the mid-1980s, recalls that ADF bashings of gay men increased in Townsville after the lifting of the ban in 1992. He says:

> Well, there were bunches of Army guys that were pissed off that they had to serve with fags now … and because the military was in all the nightclubs, and you'd have three or four AJs [Army Jerks] together, half pissed, and somebody who they would identify as being gay, whether the person was or wasn't. And they would target them and would bash them.[23]

Primary accounts indicate that servicemen and women who had established careers in non-combat roles were generally likely to find it easier to be open about their sexuality. In 1995, one serviceman reported: 'It's certainly true that some areas are still very homophobic. In some of the more traditional, macho areas—such as infantry soldiers in the army and front-line fighter pilots—homosexuality is still seen as something undesirable … in a lot of other areas, people aren't feeling so nervous about it.'[24] Cliff Anderson was working as an illustrator in the Army Survey Corps in Canberra when he made the decision to come out in early 1996. He recalls: 'Most of them were good. I mean, I even ended up taking—I had a boyfriend in February. We went to a dining-in night together, which was scary, but it was taken really well. However, there were two people that had a problem with me.' One of those people consistently left the room when Anderson entered, and the other was a female officer who made comments that gay people should not be allowed around children.[25] 'Anne' enlisted in the RAAF in 1998 and has been open about her sexuality since her recruit training ended. She has not come across any problems but observed in her early years: 'If anything, girls are more hesitant or think you might come on to them, but the guys have never been a problem.'[26]

Sandra McInerney, who served in the Army from 1986 to 2006, knew of damaging witch-hunts earlier in her career but felt she was able to be relatively open about her sexuality even before the ban was lifted. She served in the transport unit and spent seventeen years as a VIP driver, including to General Peter Cosgrove when he was CDF (2002–05). She remembers Cosgrove and his wife being very accepting of her sexuality. McInerney attained the rank of sergeant and did not feel that the policy change affected her life. She asserts, 'Did it make any difference to me? No, not at all. I was single. I was, you know, in a job, so it didn't.' However, she does remember a number of service personnel who felt able to reveal their same-sex relationships after 1992: 'And a lot of my friends who had been in for a while then felt comfortable to actually come out. [They could say], "Well, this is my partner of thirty-two years."'[27] Just as service personnel in previous decades had argued, the barrier that LGB people faced in the ADF was one of prejudice. Perceived problems such as the blackmail of homosexual personnel simply could not occur once the ADF allowed LGB people to serve openly.

Battling homophobia

Between 1992 and 2005, LGB members of the ADF had to contend with a range of continuing discriminatory policies and practices. These are apparent in submissions made to a Senate Sexual Harassment Inquiry; the unwillingness of the ADF to implement anti-homophobia programs; and the denial of partner benefits. In 1993 LGBTI activist Rodney Croome, who was then co-convenor of the Australian Council for Lesbian and Gay Rights, drew attention to the discrimination still embedded within the ADF: 'It appears that, for many people, the ADF ban was more a matter of the principle than of the welfare of actual gay men and lesbians in the ADF … Once the battle to lift the ban was won, there was little effort to follow it up.'[28] Croome was referring to one of the recommendations of the Caucus Joint Working Group on Homosexual Policy in the Australian Defence Force, which was to implement an education program within eighteen months that covered sexual orientation.[29] By mid-1993 there had been no action on this recommendation amid a rise in homophobia. An anonymous fake memo that circulated at RAAF Base Williamtown in June 1993 ordered security personnel to 'bash one poofter prior to commencement of shift … These homosexuals are known throughout the RAAF as "Fags", "Penal Poppers", "Poo Jabbers" and mostly "Gay Blades" and are affecting morale and Esprit De Corps of most Security Flight (SECFL 1) personnel.'[30]

Still, the ADF would not commit to implementing any anti-discrimination education, with ADF representative Brigadier Adrian D'Hagé quoted in the LGBTI press as describing any survey on ADF members' attitudes to be an unnecessary 'waste of taxpayer's dollars' (D'Hagé later contested that he ever made this statement).[31] Bill Bowtell, a former adviser to Paul Keating, recalls that the prime minister met one of the ADF top brass, who advised that the armed forces needed millions of dollars to fund an education program. Keating did not want to provide extra funds for such an education program, which should have been core business for the ADF.[32]

In October 1993, the office of the Minister for Defence Personnel confirmed that there would be no specific program to combat homophobia because there were already education programs about bullying and harassment.[33] Activists from outside the ADF lobbied for education to address discrimination on the grounds of sexuality. In 1993, the Australian Council for Lesbian and Gay Rights made a submission to the Department of Foreign Affairs and Trade, which was preparing a draft National Action Plan that addressed gay and lesbian issues. The council's submission pointed again to the ALP Joint Caucus Working Group's recommendation for an education program and noted the success of similar programs in the Canadian Defence Force. The council made two recommendations for the ADF: (1) 'that all ADF policies should reflect a concerted effort to improve attitudes within the ADF to lesbian and gay service personnel and prevent harassment of these personnel'; (2) 'that the ADF put in place an "Avoidance and Elimination of Sexuality Discrimination" document and incorporate it into all military training programs'.[34]

Croome recalls LGB Defence members later complaining about a sexual harassment training video:

> So, they're all sat down to watch this video about sexual harassment, and they came away feeling that the video was portraying them as the perpetrator, not the victim. So, they knew what the situation was, which was that they had been being attacked, being joked about and all the rest. But the video made it look like they're the ones that are causing the problem.[35]

Croome says that military officials with whom he met had no interest in the challenges confronting LGB personnel, so addressing these issues of discrimination and homophobia would clearly take considerable effort on a number of fronts.

Submissions to the Senate Sexual Harassment Inquiry (1994)

In September 1993, the Senate Standing Committee on Foreign Affairs, Defence and Trade commenced an inquiry into sexual harassment in the ADF. This action was prompted by allegations of sexual harassment and discrimination on several RAN ships in the 1990s, particularly HMAS *Swan*.[36] The inquiry focused primarily on gender-based discrimination against women. However, a few submissions also indicated that the ADF was ill-equipped to deal with the sexual harassment and bullying of LGB personnel.

Chief Petty Officer E.H. Sheather made a submission that linked his disapproval of women serving with gay and lesbian service. He wrote, 'The close proximity of heterosexual males and females (and homosexuals are now allowed to continue to serve in the ADF), living, working, eating, sleeping and washing together, is a recipe for disaster.'[37] T.E. Brinkley, a retired RAN commander, similarly complained about both the service of women alongside men and the service of LGB personnel. Regarding the employment of women on warships, he wrote: 'The Government, in a sop to the feminist movement, has forced on the Armed Services an unnatural and totally disliked state of affairs that was always bound to result in strife and allegations of harassment.' He continued, 'Recent pressure on the Armed Services to accept homosexuals is a parallel case which will condemn them to what they know is a badly thought-out and obviously politically motivated path.'[38] As Nicole Capezza argues, 'Sexism and homophobia are related concepts in that both stem from a gender hierarchy social structure and that similar types of individuals tend to be the oppressors of both (namely, men with traditional gender role beliefs).'[39] Sheather's and Brinkley's submissions to the inquiry affirm this by showing that many of those who struggled with the integration of women into the ADF also struggled with the inclusion of LGB people.

The Senate inquiry also received a submission from Mark Livingstone, who identified himself as a gay sergeant who had served in the RAAF for twelve years. Livingstone discussed extremely homophobic comments that had been made by Lieutenant Commander Prendergast from the Navy.[40] A letter by Prendergast, published in the *Australian Defence Journal* in 1993, included this assertion:

> In time, homosexuality will spread like a cancer through the Defence Force and morale, discipline and security problems will continue to grow as long as it remains unchecked. In the future, the fear of homosexual harassment will make the best young heterosexual Australians think twice about a career in the Defence Forces; and indeed they may be put off from doing so by homosexual officers interested in recruiting and promoting only their own kind. Parents, particularly ex-service personnel, will not wish their children to join.[41]

Livingstone stated that his submission intended to highlight 'that the incidents of sexual harassment and discriminatory behaviour within the Australian Defence Force are not confined to heterosexual incidents only' and that 'incidents of sexual harassment against homosexual service personnel are continuing whether directly or indirectly'. Livingstone also expressed his concern for gay and lesbian officers who might fall under Prendergast's command.[42]

When the secretary of the inquiry asked Prendergast to account for his comments, Prendergast stated that he had submitted his resignation from the Navy after more than thirty years of service. Prendergast asserted that 'until now I have never experienced a situation where my personal views were criticised as being contrary to my profession'.[43] Brigadier Adrian D'Hagé also responded to Livingstone's submission, asserting: 'I am on the public record as stating that education is one of the keys to effecting change in the Defence Force and ensuring the efficacy of new policy.'[44]

The final report 'Sexual Harassment in the Australian Defence Force' (1994) did little to address this problem of homophobia, devoting only two paragraphs (half a page) of the 388-page report to homosexuality. It did make one important recommendation: 'that in future, organisers of the Good Working Relationships seminars apportion appropriate time to the issue of homosexuality and the way in which the guidelines on unacceptable sexual behaviour apply'.[45] There was no further consideration of the other forms of homophobia confronting LGB members, and there was little sign that the ADF would adopt even this small recommendation.

Same-sex harassment and the lack of anti-homophobia programs

When the ban on LGB service was removed in 1992, a new Defence Instruction (General) was introduced to deal with all forms of 'unacceptable sexual behaviour'.[46] It was gender neutral and therefore should have covered both heterosexual and homosexual behaviour. The instruction included reference to actions that were prejudicial or were likely to be prejudicial to group cohesion or command relationships, which took advantage of subordinate or underage persons, and which would bring discredit to the ADF.[47] As 'unacceptable sexual behaviour' was not defined, it was open to command discretion and could be interpreted differently. As outlined earlier, LGB service members believed that programs dealing with sexual harassment portrayed gay and lesbian personnel as being more likely than their heterosexual counterparts to harass other service personnel. Rodney Croome argued in 1994 that the ADF policies 'portray gays and lesbians as sexual predators rather than people who experience harassment, violence and discrimination'.[48] As Matt Cone discovered, the instruction on 'unacceptable sexual behaviour' could be weaponised against LGB personnel.

There is evidence that other LGB service personnel were also targeted under the 'unacceptable sexual behaviour' instruction. In one case a unit commander from the Joint Movements Control Office in Sydney took action against Private Ingrid Markussen for 'embracing'. Markussen had been found hugging a fellow female soldier in her quarters at Moorebank Barracks. The charge against Markussen was dismissed, but the other soldier was charged, found guilty and given a reprimand, which would remain on her combat record. The *Sydney Star Observer* noted that the soldier would appeal this decision. Markussen told the *Sydney Star Observer* that she had been singled out for abuse within the ADF because she was a lesbian and that she had suffered 'offensive sexual comments and abuse' since April 1996. She also described an incident in September 1996 where a corporal in her unit allegedly told her: 'All gays should be shipped to Tasmania, lined up and shot.' She complained about the statement, but her complaint was dealt with in a single mediation session before being 'swept under the carpet'. She described a litany of other continuous harassment, which she said had resulted in her breaking down and being comforted with a hug from another soldier. This was the incident for which Markussen was charged. Not surprisingly, after her case was heard, Markussen applied for discharge from the Army.[49]

Although Brigadier D'Hagé had affirmed his belief in 1994 that education was an important step in ensuring the full inclusion of LGB service personnel, such policies were not rolled out in a sustained way until 2000. In 1994, an ADF representative asserted that the Navy's Good Working Relationships program addressed homosexuality but stated that the Army and RAAF did not have equivalent programs. It is worth stressing that the Good Working Relationships program had been developed to address the sexual harassment of women. Rather than educating service personnel about homosexuality, it appears that homosexuality was discussed as part of the 'unacceptable sexual behaviour' section.[50] The ADF representative stated that there were no other specific programs that addressed homophobia.[51] In 1994, an Air Force officer spoke out anonymously, arguing that tri-service educational programs were important because they had the potential to make a real different in changing homophobic attitudes. He stated that 'a lot of people in the Air Force don't know anything about homosexuality, except what they were taught as kids' and that 'it would be nice if something could be done about that, just to make people open their minds and realise things have changed'.[52]

LGB advocates within the ADF argued strongly for programs but pointed to a lack of willingness on the part of ADF officials to engage with them. In 1996 David Mitchell, who was speaking in his capacity as president of G-Force—a support group for LGB Defence members discussed below—asserted, 'When I formed G-Force, I thought it might be a vehicle to coordinate an education program but they weren't really committed to the idea and it got knocked right back.'[53] The courage of ADF personnel such as Mitchell to make such statements deserves particular recognition. The ADF in this period was still grappling with the treatment of LGB personnel, and being identified as an openly gay person in the service could result in discrimination. Furthermore, the ADF more generally discouraged (and still discourages) service personnel from publicly challenging ADF policy. Yet making such a statement was critical to show that LGB people were serving and that they were working for change.

By 2000 the ADF was finally running a structured educational module to address homophobia. In April 2000, the *Melbourne Star Observer* reported that the ADF had been 'quietly taking advice' from a New Zealand company, Full Spectrum Ltd, on how to change its culture 'to be more "gay friendly"'. Full Spectrum Ltd's director Eugene Moore asserted: 'The fact of the matter is that gays and lesbians are "coming out" in civil society. And organisations, whether military or not, have to create an environment where those people who are out in the rest of their lives don't have to go back into the closet at work.'[54] The outcome of consultations with Moore was the design of an optional online training module entitled 'Understanding homosexuality'. It was not faultless, but it was a first step towards a tri-service education program to combat homophobia.

In August 2000, a widely reported incident showed the need for such educational programs. ADF peacekeepers in Bougainville made a tasteless 'Mardi Gras' video showing 'mostly male defence personnel dressed as large-breasted women on military vehicles to a backing soundtrack [of] "Freak" by the band Silverchair'.[55] An internal investigation concluded: 'A few of the participants in the parade and the social event that followed tested the boundaries of accepted good taste with their behaviour. Some behaviour was clearly offensive.' Yet both the investigator and brigadier commanding the Peace Monitoring Group described the incident as soldiers who had been working hard for several months who were just having a bit of fun, so they recommended no disciplinary action.[56]

For LGB Defence members and activists, though, the incident was reflective of wider, structural homophobia within the forces. Rodney Croome asserted that the video was an indictment of the ADF and government, which refused to acknowledge the damage and danger of prejudices against minorities: 'We hear a lot about the dangers of not upgrading our military hardware but there are just as many dangers in not upgrading the attitudes of military personnel.'[57] Clearly, greater efforts were necessary for Defence members genuinely to understand sexuality and how their behaviour could affect LGB members.

HIV and AIDS

Throughout the 1980s and early 1990s, HIV and AIDS were often conflated with homosexuality. As Deborah Lupton has argued, 'the arrival of AIDS, and the linking of male homosexual behaviour to its aetiology' served to demonstrate 'the thin layer of acceptance which masks deeper anxiety and negative reactions to sexual difference'.[58] This considered, it is perhaps not surprising that opponents to LGB service utilised the issue of HIV and AIDS to argue against the inclusion of LGB service personnel. LGB advocates argued against discrimination on the grounds of HIV-positive status, but their efforts were not successful.[59]

In March 1987, the ADF indicated that it would not screen recruits for HIV. Later in the year, in a Senate Estimates Committee hearing, Defence representatives revealed that several men had been discharged for being HIV positive. RAAF Police had used coercive measures including secret tests and contact tracing, and this prompted the Defence Minister to direct the ADF to come up with a formal policy on HIV and AIDS.[60] DI(G) Pers 16-6, promulgated in March 1988, authorised HIV prevention education programs across the services and counselling for anyone tested for HIV (especially those who tested positive) and contact tracing. The most important parts of the policy, which set precedents for the next thirty years, related to managing people who tested positive for HIV. Any new enlistee would be tested upon commencement at basic training; those who tested positive would have to discharge. Those who seroconvert (contract HIV) while serving would not be discharged from the ADF unless diagnosed with a category 3 or 4 HIV infection—defined as enlarged nymph nodes and the emergence of other diseases and opportunistic infections respectively. Persons who were in the asymptomatic categories 1 and 2 were 'subject to a formal medical board and their future employability is to be determined on a case by case basis on the advice of the respective Director General Health Services. Group II HIV infection should not in itself be sufficient grounds for discharge … Category of employment, as well as medical and physical fitness for duty, are to be considered.'[61] Because HIV was seen as a proxy for homosexuality, though, being HIV positive usually did lead to discharge before 1992. As explained in more detail in chapter 7, only in 2018 did the ADF update this policy to provide more flexibility for serving members living with HIV.

The ADF was aware that several service personnel had been diagnosed as HIV positive. In November 1992, Defence representatives advised a Senate Estimates committee that thirty-four Defence members had tested HIV positive between 1988 and 1992. Of this figure, twenty individuals had been discharged from the ADF.[62] In 1993, the Australian HIV Surveillance Report from the National Centre in HIV Epidemiology and Clinical Research showed that thirty-six of the 66 380 ADF personnel were HIV positive. Of this figure, thirty-five were male and one was female. More than half of these people had contracted HIV through heterosexual contact. It was reported that all thirty-six had been removed from combat duty.[63] While this data showed that the majority of cases of HIV transmission were through heterosexual contact, it did not stop the chief of the RSL, Major General Digger James, from arguing the following year that homosexuals 'could pass AIDS to wounded soldiers during direct blood transfusions in the field'.[64]

In October 1994, the Australian Federation of AIDS Organisations (AFAO) called for a review of the ADF's policy on HIV and AIDS. It cited a case where a serviceman working in electronic warfare tested positive to HIV after an overseas posting. The serviceman was told he was no longer eligible for further postings or promotion. AFAO's president, Tony Keenan, asserted that 'the Defence Force policy is inconsistent with the Government's generally enlightened policies on HIV/AIDS. More to the point, the policy is in clear breach of the Disability Discrimination Act.'[65] The serviceman lodged a challenge with the Human Rights and Equal Opportunity Commission (HREOC). It was one of a number of such complaints the HREOC received that year.[66] In response to the publicity this received, Brigadier D'Hagé asserted that the presence of HIV-positive people on a RAN ship was a threat to morale and introduced the possibility of HIV-contaminated blood on the battlefield. He continued, 'If there is a missile strike, we've got a problem with having HIV-positive people on board.'[67]

For some recruits who tested HIV positive, this was the first time they found out their status. In 1995 new recruit Wayne Oldrey learned that he was HIV positive two weeks after his blood test, and the ADF health officials did not manage the situation well. He recalls, 'They were in space suits, I kid you not. They were in the full biohazard suits.' This hysterical overreaction was after more than a decade of HIV research showed that the virus could not be transmitted casually. The treatment meted out to Oldrey only compounded the distress of receiving the HIV-positive diagnosis. In line with the ADF policy on HIV, Oldrey was medically discharged from the Army.[68]

In July 1995, another HIV-positive recruit (given the pseudonym 'Mr X') who had been dismissed from the ADF in 1993 filed a complaint in the HREOC for breach of the *Disability Discrimination Act 1992* (Cwlth), which included protections for people living with HIV. The HREOC Inquiry Commissioner found that 'the complaint has been substantiated and the dismissal of the complainant from the ADF on the grounds he was HIV positive were unlawful'.[69] In response to the ruling, as one of its final acts the Keating Labor government issued regulations under the Disability Discrimination Act that exempted the ADF and allowed it to continue the practice of banning recruits who tested positive for HIV. The Australian Medical Association slammed the move, with its vice president, Dr Keith Woollard, stating: 'HIV is an infection, not a disease ... It does not impair anybody's ability to perform a particular task.'[70]

Meanwhile, the ADF did not accept the HREOC decision and took the matter to court. In October 1996, a Federal Court judge concurred with the HREOC determination.[71] The ADF was successful, though, on further appeal: in 1998, the Full Bench of the Federal Court overturned the earlier ruling, upholding the ADF's right to dismiss the soldier who was HIV positive on the grounds that the 'inherent requirements of the job' exempted the ADF from HIV-related protections under the Disability Discrimination Act.[72] Mr X then appealed to the High Court of Australia, which, by a 5–1 majority, upheld the Full Federal Court decision.[73] Alison Young argues that the judicial discourse surrounding this judgement was highly problematic. She contends:

> The army is configured as a fraternal order, entrusted with the safety of the nation and trusting in each other. That mutual fraternal trust is endangered by the spectre of the HIV-positive man whose blood cannot be used as the life-giving transfusion for the injured and whose blood may put at risk his fellow soldiers. His status, as the embodiment of infection, means that he can no longer be part of the fellowship of soldiers; he must be put out of its ranks.[74]

After the High Court judgement, the ADF framework for managing HIV and AIDS remained relatively unchanged until 2018.

Although ADF policies left scope for those who seroconverted to continue serving, this was quite difficult in practice. 'Nelson' was our only interview participant who served during the 1990s and 2000s while HIV positive, and he kept this secret for the entire period of his service. Nelson had joined the RAAF in the 1980s and was diagnosed as HIV positive in the mid-1990s. He says that the diagnosis 'frightened' him not only from a health perspective but also with regard to his career. Nelson says that the practice for managing service members living with HIV 'was basically instant discharge because you couldn't serve, you couldn't go to war, so to speak, because you had this disease and your mates couldn't rely on you. And if you were injured even through just a practice exercise or more of a real wartime, no one would want to care for you because you were HIV positive. So it was an instant discharge.'

Nelson kept his diagnosis a secret, and he found a fortunate ally in his medical officer, who did not disclose Nelson's diagnosis. When he was due for routine blood tests, the doctor would concoct reasons to defer Nelson's tests. After about five years it reached a point where the doctor could not defer any longer. The doctor did a blood test but, as Nelson recalls, 'The doctor said to me: "You will never ask me any more questions about this again." That's where it ended for me. I don't know what happened to those blood tests ... However, I lived to tell the story that I got to serve twenty-one years in uniform.'[75]

Nelson became adept at hiding his HIV status from his friends and colleagues. Nelson was fortunate that in 1996 the advent of Highly Affective Anti-Retrovitral Therapy (HAART) transformed the landscape for people living with HIV: it was now possible to live long and healthy lives. In the early years, HAART was a cocktail of medications that had to be taken at specific times, some of the pills with meals and others without food. Nelson had to adjust his schedule around his medication, which affected his social life. He hid his pills and found workarounds when attending courses or weekend functions such as barbeques.[76] As chapter 7 relates, the experiences of service members living with HIV still leave much to be desired, and the ADF is only now revising its policies to align with the science around HIV.

Ongoing structural discrimination: denied benefits for same-sex couples

One of the most hard-fought LGB reforms in this period was the recognition of same-sex relationships by the ADF. As early as 1993 two lesbian members applied for the Army to recognise their relationship; the Army rejected their application because the 1986 administrative order on 'Recognition of a Person as Family' defined a de facto spouse as 'a person of the opposite sex who, although not legally married to the member, lives with the member on a permanent and bona fide basis as the member's spouse'.[77]

There are important practical reasons behind the provision of military support to Defence members' partners. The higher risks associated with military service and the total nature of the commitment mean that the ADF needs to provide particular forms of support to the families of service personnel, who often make considerable personal and professional sacrifices to support the ADF member's engagement. For instance, ADF personnel are expected to relocate more frequently than workers in civilian occupations. The geographical relocation (posting) of ADF members and their families to different parts of Australia and sometimes overseas is required to maintain the operational effectiveness of the ADF.[78] However, such relocation can pose challenges to the family of the ADF personnel, including social isolation and difficulties in securing new employment.[79] As a result, the ADF provides extra financial support to account for the partners and families of the ADF service member.

The exclusion of same-sex partners from ADF consideration and compensation placed particular hardship on those couples.

Many LGB service personnel were directly affected by policies that discriminated against them. Keith Jeffers, himself an ex-airman, remembers how hard it was for him and his partner Mike Seah, a doctor in the RAAF:

> We had everything together and it was just really about if something happened to each other. I don't want to be in a hospital room over his body arguing with his parents. I didn't want to do that. I think that was what was the worst part. It wasn't necessarily the money or anything else. It was just logistical kind of stuff and it just didn't seem fair.[80]

For some LGB Defence members, this situation was intolerable. In 2003, Captain Rebekah Beaumont wrote that she had witnessed dedicated soldiers leave because of the inequity: 'I have seen them become increasingly frustrated about not having the same entitlements as their heterosexual counterparts and even the demise of their relationships because of the lack of support and entitlements.'[81]

The failure to recognise same-sex couples affected not only active service members but also veterans. In 2003 the United Nations Human Rights Council (UNHRC) found that Australia had violated its international obligations by denying Edward Young, the surviving partner of a World War II veteran, a widow's pension on the basis of his sex or sexual orientation. Young, who had been with his partner for thirty-eight years and had nursed him for the last five years of his life, had applied for a veteran's pension after his partner died in December 1998. This was an entitlement that was available to the heterosexual partners of service personnel who served overseas but was denied to Young because an interdependent was defined as someone of the opposite sex.[82] Young wrote to the *Melbourne Star Observer*: 'The Government's unwillingness to grant me the same rights and benefits it automatically grants to the heterosexual spouses of former service personnel has been very distressing. I feel it is an insult to the memory and war service of my partner.'[83] Unfortunately, the Howard coalition government dismissed the UNHRC ruling and would not extend veterans' benefits to same-sex couples. That reform was finally implemented in 2009 after the Rudd government amended eighty-five pieces of legislation to recognise same-sex de facto relationships.

The push for equity and to extend benefits to same-sex partners was a sustained one with many obstacles. It occurred primarily in two stages. The first stage was between 1994 and 1997 and was led by the advocacy group

G-Force. The second stage occurred between 2002 and 2005 and was driven by the successor to G-Force, DEFGLIS.

G-Force

G-Force was established in June 1994 by RAAF Sergeant David Mitchell. Mitchell's work background provided him with advocacy experience. In 1993 he had been posted to the United Kingdom to work with the Royal Air Force on Exercise Long Look, which was a professional development exchange between members of the Australian and UK armed forces. Mitchell was open about his homosexuality, but the United Kingdom did not lift its ban on LGB service until 2000. Mitchell knew that the issue of LGB service was generating considerable interest in the United Kingdom. Consequently, when he was asked if he would speak publicly on the television program *Taking Liberties* about LGB service in the ADF, he agreed to do so.[84] Mitchell was permitted to speak publicly on this issue and appear in uniform. However, the letter from the Defence attaché at the Australian High Commission warned that 'any comments made by him during the interview must be considered as his personal views. Sergeant Mitchell was advised to think very carefully before agreeing to take part in the documentary, purely for the protection of his own privacy, but no restrictions were placed upon him.'[85] The UK Defence establishment later advised the ADF that they would prefer Mitchell not be sent back to the United Kingdom on future exchanges.

The issue of Australian LGB Defence members on exchange resurfaced a few times. In 1995 an airman on exchange to the United Kingdom was gay. Although this sergeant was two and half months into his four-month posting, he was still sent off Shawbury Base with just twenty-four hours' notice.[86] In 2000 another gay RAAF officer, Mike Seah, was posted to the United Kingdom to undertake a Diploma of Aviation Medicine. Seah was initially told he was selected for the post, but not approved because of his sexuality. On threat of resignation, the RAAF reversed its decision and allowed the posting to proceed. However, Seah was advised that approval was conditional on him being discreet about his sexuality. Ultimately this turned out to be unnecessary: the UK ban on LGB service was lifted less than a week after he arrived.[87] David Mitchell was asked for media comment on the 1995 case; by that stage, he was able to provide it in his role as president of an organisation for LGB members of the ADF known as G-Force.

Mitchell remembers that the genesis for G-Force came from his own difficult coming out, as well as from his experience in the United Kingdom. Not only had he showed great courage in advocating publicly for the rights of LGB service personnel, but also he had formed contacts with Rank Outsiders, a support group for British LGB service personnel.[88] G-Force was formally established in June 1994.[89] Mitchell and friends, most notably RAAF airman Stuie Watson, drafted a constitution and set out the aims of the organisation. Membership was open to any LGB current or ex-service person. G-Force's first press release noted that, 'as coming out can be a painful and lonely experience, particularly in the military environment, a need has been identified for a support group to be established to enable networking and exchange of experiences and support'.[90] On establishment, G-Force listed seven objectives:

1 Provide support and guidance for gay servicemen and women—without infringing service laws.
2 Provide support and guidance for gay ex-servicemen and women—without infringing service laws.
3 Make contact with other national and international organisations involved in service life.
4 Collect information, from any source, that is relevant to the improvement of gay rights within the Australian Defence Force.
5 Encourage the Department of Defence to enhance its attitude towards homosexuality within the Defence Force.
6 Make contact with other national and international organisations, so as to encourage awareness and education regarding Gay lifestyle and their existence in the Australian Defence Forces.
7 Make contact with other national and international organisations, to participate and be involved in Gay community events.[91]

Mitchell was keen to emphasise the important social role that G-Force could play. He told *Navy News* that he hoped 'that through this group gay service personnel will band together to form a strong network of support to help lessen the trauma felt by others as they go through this journey'.[92]

G-Force eventually accrued a membership of about thirty-five people. Members Mike Seah and Karl Bryant remember that the group drew from across the three services, but there was a proliferation of RAAF members, given the organisation's leadership. The organisation held meetings in Sydney but still tried to reach out across the country. Bryant recalls working for a supportive RAAF officer who actually encouraged him to fly to Sydney and join G-Force. Bryant became the main Melbourne contact, so he met several gay and lesbian Defence members. He explains:

> I had a few people, lesbians and young gay men, who spoke to me. But unfortunately they were so scared about what could happen to them, even though they knew where I'd gone and been

> and [was] continuing to head, and the group, the G-Force, that we had formed—they were glad that they could talk to someone about it, but they were never going to stay in the Air Force.[93]

The ADF hierarchy was ambivalent about G-Force, and the group sometimes received homophobic letters in its post office box. These included news clippings from the ADF's Christian newsletter, photos of naked women, a picture of a grave, clippings from pornographic magazines and notes with statements like 'Homosexuals can change'. Only once did Mitchell receive an explicitly threatening letter: a cartoon with a character named David racing towards a cliff and the caption 'Will Dave survive tomorrow?'[94] It was this very sort of homophobia and the experiences Mitchell, Bryant and other G-Force members heard from across the ADF that made the organisation so important.

The struggle for couples' recognition was the other major concern for G-Force. During the eighteen-month period to February 1995, there were six Army applications for same-sex de facto recognition, all rejected. G-Force faced an uphill battle to garner support from straight service members and the top brass. In 1995 the Army Soldier Attitude and Opinion Survey asked respondents whether they agreed with the statement: 'The Army should recognise homosexual couples as being in de facto relationships.' The results are compiled in table 6.[95]

Table 6: Responses to question on recognising same-sex de facto relationships in Army Soldier Attitude and Opinion Survey, 1995

Group surveyed	Strongly agree (%)	Agree (%)	Undecided (%)	Disagree (%)	Strongly disagree (%)
Officers	3	7	16	18	56
Other ranks	1	5	14	15	65

G-Force members Karl Bryant and Mike Seah both lodged applications to recognise their same-sex partners. Bryant even had the support of his commanding officer and advice from a legal officer, who observed that the rule 'does not appear to reflect the policy statement of 1992 proclaiming an end to discrimination in the ADF on the basis of sexual preference'.[96] When both Bryant and Seah's applications were rejected, they lodged redresses of grievance. At each level of appeal the RAAF consistently rejected their complaints on the grounds that the existing policy was legal and there were no plans to change it. The Chief of Air Staff advised Bryant that 'should Government endorse changes to the PSA [Public Service Act], Headquarters Australian Defence Force will study the possible implications for members of the Australian Defence Force (ADF)'.[97] Seah appealed his redress of grievance as far as CDF General John Baker, who responded: 'While you argue that the common [social] standard requires that the discrimination you complain of should be removed, I do not believe that common standards make such a demand.'[98]

Unsuccessful within the ADF, like earlier LGB Defence members, both Bryant and Seah took their cases to outside bodies. Bryant lodged a complaint with the Defence Force Ombudsman, but this proved fruitless; it very quickly informed Bryant that it could not investigate his complaint and suggested he complain to the HREOC. Bryant took that opportunity, but unfortunately the outcome was not favourable. It was almost three years before the HREOC forwarded Bryant the Australian Government Solicitor's advice: 'As none of the definitions of "spouse", "family", and "de facto" referred to above are capable at law of applying to a same-sex partner, a denial of any allowances to FSGT Bryant (if in fact he would otherwise have been entitled) on that basis could not constitute an act or practice of discrimination.'[99] Mike Seah tried multiple routes: he lodged a submission to the 1996 Senate Inquiry into Sexuality Discrimination; unsuccessfully applied for the ADF to recognise his partner as a family dependant; and sent letters to the Minister for Defence Industry, Science and Personnel, Bronwyn Bishop. All of these efforts came to nought.[100]

Although these G-Force members did not achieve the reforms to partner recognition that they desired, the organisation successfully advocated for some gains for same-sex partners. Mitchell recalls working with the chair of the Army Health Benefit Scheme (now Defence Health) to change its rules to permit same-sex partners to access member benefits. G-Force was also successful in gaining the right for same-sex partners to access spouse security passes for base admissions.[101] There was only one other recognition for same-sex couples: in 2000, Senator Jocelyn Newman, Minister Assisting the Minister of Defence, confirmed: 'A serving Australian Defence Force (ADF) member may nominate a same-sex partner as their next of kin for casualty notification purposes only. The same-sex partner would be notified in the event of the ADF member being the subject of a casualty report.'[102] With no change in sight, securing the full remit of recognition for same-sex couples was going to take a new generation of LGB activists within the ADF.

Mardi Gras

In 1996, G-Force significantly increased the visibility of LGB service personnel when they marched in that year's Sydney Gay and Lesbian Mardi Gras parade. In many ways, Mardi Gras participation is emblematic of LGB service in this period. G-Force had to advocate strongly for approval to participate in Mardi Gras and was permitted to do so only if participants marched in plain clothes. Displays of LGB pride had to be discreet and involved a substantial amount of work on the part of those involved.

David Mitchell had first raised this idea in late August 1995, when he wrote to Brigadier Adrian D'Hagé advising that G-Force intended to march in Mardi Gras in 1996. The letter stated: 'We [G-Force] are becoming a significant group within the Australian gay community and as such we feel we have the right to express ourselves in celebration of this position … they [Mardi Gras Association] are to affirm the pride, joy, dignity and identity of our community and its people.'[103] Mitchell also believed that participation in Mardi Gras would play a valuable role by showing other LGB Defence members watching the parade that G-Force existed as a support service.[104] Indeed, it would be the Mardi Gras announcement that first alerted Mike Seah to G-Force's existence and motivated him to join.[105]

D'Hagé initially supported Mitchell's request. By November 1995, however, the ADF position had changed. The Vice Chief of the Defence Force, Vice-Admiral Robert Walls, prohibited G-Force from participating in Mardi Gras. D'Hagé stated to the media:

> When we look at the charter of the Gay and Lesbian Mardis [*sic*] Gras Pty Ltd, it is not only a celebration of their sexuality, it is also a promotion … The difficulty we have with that is that it is very doubtful as to whether an official Defence Force float is an appropriate vehicle to promote a particular lifestyle. We would make the same decision if they were participating in a heterosexual parade.[106]

While Defence argued that it was not discriminating in preventing G-Force from marching, D'Hagé remembers an admiral telling him that the float would go ahead 'Over my dead body'.[107] Mitchell remembers a senior officer ringing him and saying, 'Well, if you go ahead with this, I will ruin you.' Mitchell hung up on him.[108] Mitchell and D'Hagé knew that the ADF could not legally prevent G-Force members from marching in the parade in civilian attire. D'Hagé states that he spent the next month working behind the scenes with the CDF, arguing: 'If you don't let this damn thing in, you know your policy of treating people fair and equitable will be out the door. This will become—the non-approval will become the story.'[109] In January 1996 the press reported that the ADF had agreed to support a G-Force float in the Mardi Gras as long as it was 'low key' and 'non-military' in appearance.[110]

Mitchell stated that the float was important to change Australians' perceptions of gay people: 'It allows the community to see that gay men and lesbians are from all walks of life. The aim is to thrust our group into the spotlight to offer support and show gay and lesbian people in the forces that there are others.'[111] The RSL condemned the ADF's decision, its national president, Major-General Digger James, declaring: 'I see it really as defence force bosses proselytising on behalf of a sectional group.'[112] Such statements showed just how important it was for G-Force to march in Mardi Gras, proudly making public LGB service, which had been kept silent for far too long.

Between ten and twenty serving members and their partners marched in the G-Force float.[113] In the ABC broadcast of the parade, host David Marr joked: 'And teamed up with Leather Pride, I think they could probably defend the country.'[114] David Mitchell recalls: 'We were being able to be—to use that expression "out and proud" in front of the world's media and people in our own country and say that, well, gay and lesbian people want to serve and serve our country the same as straight people.'[115] Mike Seah's mother did not approve of him marching in the Mardi Gras—so instead he drove the truck carrying the G-Force float. He remembers the importance of the participation in Mardi Gras: 'It was nice to actually have that visibility ... although being gay and lesbian was OK in the Defence Force for the previous four years there was not really a lot of visibility about, so it was the first step towards some sort of recognition that there were gays and lesbians in the military.'[116]

The proud participation of ADF service personnel in the country's most prominent LGBTI event was ground-breaking. It positioned those who participated both as LGB people and as military personnel, claiming and displaying pride in both of those parts of their identities. It also showed both LGBTI people and the broader community that Defence personnel could be members of Australia's LGB community. G-Force did not participate again in Mardi Gras, largely owing to logistical and financial challenges. Karl Bryant, though, participated in a much smaller group of service members who, against the wishes of the CDF and other senior officers, marched in 1999 under the banner Defence in Unity.[117] G-Force continued as an organisation until about 1998, when the financial and mental toll of running the organisation led Mitchell and other organisers to disband it. Having laid the foundations for important reform, G-Force showed that it was possible to organise and advocate for real change to support LGB Defence members. It would take a new generation of leaders to advance the equality agenda in the 2000s.

DEFGLIS

DEFGLIS—the Defence Gay and Lesbian Information Service (renamed Defence LGBTI Information Service in 2011)—was founded in 2002, and it continued the advocacy work G-Force had undertaken. The founder of

DEFGLIS, Stuart O'Brien, was a petty officer in the Navy serving in Canberra and had already been advocating within the ADF for the recognition of same-sex partners. He remembers that discrimination affected him in a number of ways, including during the process of relocating and in the reduction in the accommodation allowance he received. He explains: 'So, because he [my partner] was a civilian, I only got half the housing benefits so I could only get rental assistance for one person sharing with another person, so we had to pay more for our rent.'

O'Brien founded DEFGLIS with the aim of providing information and referral services to LGB members of the ADF. O'Brien remembers, 'It was just an email newsletter that kicked off and just started to spread and people were starting to get it. And we'd just go out with some health information or "This is what's in your state or territory" and that was really the creation of it.'[118] The email distribution list was kept private, although O'Brien courageously displayed his photograph and a contact email on a Geocities webpage.

Deployment overseas—particularly in times of conflict—highlighted the need for same-sex partner recognition and kept the issue on the public agenda. In 1999 Australia led the multinational peacekeeping taskforce in East Timor, International Force East Timor (INTERFET).[119] From 2001 and 2003 respectively, the Australian military was engaged in the Afghanistan and Iraq conflicts. For deploying members, the issue of partner recognition took on grave significance. Keith Jeffers remembers that when Mike Seah deployed to East Timor:

> It was just really difficult for me to … or for us to understand what would happen if something happened to either of us. It's possibly less with me because I was obviously in the country for most of it. He was on the first plane into East Timor and that really scared me. Sorry … I went out that night with some friends because he'd left and [someone said], 'Oh, you're a war widow again.' I was really, really scared and I just thought it's not fair.[120]

This overseas engagement underlined for politicians the need for LGB partner recognition in the ADF. While troops were in East Timor in 1999, Andrew Bartlett from the Australian Democrats, a long-time advocate for same-sex couple recognition, took the opportunity to draw attention to the lack of spousal rights. He stated: 'It is bad enough that citizens who pay taxes like everyone else don't get the same recognition of their relationships, but when it is service personnel who are risking their lives overseas and the government refuses to recognise their nominated loved ones—that is outrageous.'[121] In 2003 openly gay Democrats Senator Brian Greig consistently raised the issue of recognition for same-sex couples by the ADF. That year the ALP joined the Democrats in endorsing same-sex relationship recognition.[122]

An FOI request to recover the paper trail over the ADF's deliberations on de facto recognition turned up nothing, so O'Brien's oral history interview is the best available source to reconstruct what ultimately led to policy change. O'Brien attended a presentation sometime in 2001 or 2002 by the Sex Discrimination Commissioner, Pru Goward. O'Brien asked whether she was aware that the ADF would not recognise same-sex couples and that they were justifying it on the basis of a definition within the Sex Discrimination Act. A few months later, Goward's office contacted O'Brien with an important piece of information: the ADF could set whatever definition of de facto spouse it wished. There was no need to wait for the public service or for the government to intervene. O'Brien subsequently set up DEFGLIS and began to lobby through the ADF chain of command.[123]

It is not clear what led the ADF finally to amend its policy, but there was one change that O'Brien believes was critical: the appointment of Air Chief Marshal Angus Houston as CDF in July 2005. Sometime in 2004 O'Brien wrote a letter to CDF General Peter Cosgrove but did not receive a favourable response: Cosgrove advised that he would not change the policy unless the government ordered him to do so. Houston, however, clearly did not feel the need to wait for the Australian Government. In October 2005, Defence flagged the impending recognition of same-sex relationships. The new instruction DI(G) PERS 53-1 was released on 1 December 2005 and redefined an interdependent relationship as 'a person who, regardless of gender, is living in a common household with the member in a bona fide, domestic, interdependent partnership, although not legally married to the member. This … also allows for those now recognised on the basis that they are in an interdependent same-sex partnership.'[124] A representative of the ADF stated, 'ADF members of the same sex in recognised interdependent relationships will receive the same range of defence conditions of service as those members in ADF recognised de facto relationships.' Stuart O'Brien told the media that 'It's a huge step forward'. He also stated that he was being posted to Iraq in December and was reassured to know that his partner would be financially secure if something were to happen to him.[125]

An indication of the distance traversed in this period towards LGB inclusion can be gleaned from the reaction from the national president of the RSL, Major-General Bill Crews. In a considerable shift from previous RSL statements regarding LGB rights, Crews stated that 'the RSL would support a gender-neutral provision for allowances to support those who have … same-sex relationships'.[126]

Conclusion

The ban on lesbian, gay and bisexual service in the ADF ended in November 1992. By December 2005, the ADF had extended partner recognition to same-sex couples. Other policies that disadvantaged LGB service personnel, such as a lack of diversity education programs, were successfully challenged during those thirteen years, and support groups such as G-Force and DEFGLIS provided important visibility and forums for LGB advocacy. While restrictive policies in relation to HIV and AIDS remained unchanged, the position and service of HIV-positive personnel did receive some public attention. The degree to which LGB service personnel felt able to be open about their sexuality varied throughout this period. Some LGB personnel found that they were able to disclose their sexuality and were accepted by their peers and superior officers. Others found that tolerance had its limits or that they faced discrimination that could be severe.

The changes achieved in the period 1993–2005 were substantial. LGB personnel had long been a part of the Australian military but, finally, this was public knowledge and not something that service personnel had to hide. Moreover, the ADF's public statement when it extended partner benefits to same-sex couples in December 2005 noted: 'Defence places great emphasis on ensuring its people work in an environment that is fair and inclusive, recognising that this enhances operational capability and effectiveness.'[127] This use of the word 'inclusive' implied an important shift in the organisation's approach to sexuality. However, there were still considerable reforms that needed to be made for LGBTI individuals. In particular, as chapter 6 outlines, the position of transgender and intersex service personnel remained unaddressed. It would take a different set of brave service members in the 2000s to challenge the ADF to include them as well.

Notes

1 Captain J.S. Salter, *2/95 Soldier Attitude and Opinion Survey: Equal Employment Opportunity Issues* (Canberra: 1st Psychological Research Unit, Australian Army, May 1996), pp. 12–14; *1995 Officer Attitude and Opinion Survey: Equal Employment Opportunity Issues* (Canberra: 1st Psychological Research Unit, Australian Army, February 1996), pp. 7–9; *1995 Soldier Attitude and Opinion Survey: Equal Employment Opportunity Issues* (Canberra: 1st Psychological Research Unit, Australian Army, February 1996), pp. 7–9; Major U.J.A. Weisse, *Report on the Officer Attitude and Opinion Survey: Topical Items 1993–94* (Canberra: 1st Psychological Research Unit, Australian Army, June 1995), p. 25; Captain J.S. Salter and Captain S.E. Hodson, *Soldier Attitude and Opinion Survey: Topical Items—1992* (Canberra: 1st Psychological Research Unit, Australian Army, September 1995), pp. 5–6.

2 Martin Chulov, 'Gays should not be in the Army—RSL', *Australian*, 4 March 2000, p. 7.

3 Wadham, 'Brotherhood', p. 226.

4 Ibid., p. 229.

5 Ed Bailey, interview.
6 David Wells, interview with Noah Riseman, 12 August 2015, Canberra.
7 'Jason', interview.
8 Ed Bailey, interview.
9 'Pencil', interview.
10 Kate Carlisle, interview.
11 'Pencil', interview.
12 'Kevin', interview.
13 'Bill', interview.
14 'Kevin', interview.
15 'Bill', interview.
16 Kayt Arthur, 'Gay officers must come out', *Melbourne Star Observer*, 25 August 1995, p. 3.
17 Matt Cone, interview. See also Riseman, Robinson and Willett, *Serving in Silence?*, p. 204.
18 Riseman, Robinson and Willett, *Serving in Silence?*, pp. 200–19.
19 Paul Chamberlain, 'Gration on gays, war and peace', *Sydney Morning Herald*, 16 April 1993, p. 4.
20 David Morrison, interview with Noah Riseman, 19 June 2017, Canberra.
21 Aaron Belkin and Jason McNichol, 'The effect of including gay and lesbian soldiers in the Australian Defence Forces: Appraising the evidence', Center for the Studies of Sexual Minorities in the Military, University of California, 19 September 2000, p. 31, http://www.dont.law.stanford.edu/wp-content/uploads/2010/11/Australian_Defence_Forces.pdf, retrieved 10 January 2020.
22 Vince Chong, interview with Noah Riseman, 16 November 2016, Canberra. See also Riseman, Robinson and Willett, *Serving in Silence?*, p. 245.
23 Colin Edwards, interview.
24 Martyn Goddard, 'Forcing the issue', *Sydney Star Observer*, 20 April 1995, p. 10.
25 Cliff Anderson, interview with Noah Riseman, 24 August 2017, Brisbane.
26 'Anne', interview with Noah Riseman, 24 July 2015, Melbourne.
27 Sandra McInerney, interview.
28 'ADF gays stay in the closet', *Melbourne Star Observer*, 15 October 1993, p. 4.
29 *Report of the Caucus Joint Working Group on Homosexual Policy in the Australian Defence Force (ADF)*, p. 7.
30 Barbara Farrelly, 'Air Force to probe gay hate', *Sydney Star Observer*, 25 June 1993, p. 3.
31 Mark Forrest and Peter O'Shea, 'ADF chief: Gay survey is "a waste of money"', *Capital Q*, 9 July 1993, p. 1.
32 Bill Bowtell, interview with Noah Riseman, 26 May 2017, Sydney.
33 'ADF gays stay in closet', p. 4.
34 Australian Council for Lesbian and Gay Rights, 'Protecting and promoting lesbian and gay rights: A submission on the draft National Action Plan', 11 November 1993, p. 11, NAA A9737, 1993/60684 PART 1.
35 Rodney Croome, interview.
36 Kathryn Spurling, '1991–2001: The era of defence reform', in *The Royal Australian Navy*, pp. 281–2.
37 E.H. Sheather, 'Submission to the Standing Committee on Foreign Affairs, Defence and Trade: Reference on sexual harassment in the Australian Defence Force', 15 November 1993, p. 5.
38 T.E. Brinkley, 'Submission to the Standing Committee on Foreign Affairs, Defence and Trade: Reference on sexual harassment in the Australian Defence Force', 8 November 1993, p. 70.

39 Nicole M. Capezza, 'Homophobia and sexism: The pros and cons to an integrative approach', *Integrative Psychological and Behavioral Science* 41, no. 3 (2007): 248.
40 M.L. Livingstone, 'Submission to the Standing Committee on Foreign Affairs, Defence and Trade: Reference on sexual harassment in the Australian Defence Force', 23 January 1994, p. 577.
41 Gavin Prendergast, 'Unacceptable sexual behaviour—the real victims', *Australian Defence Force Journal*, no. 102 (September–October 1993), n.p.
42 Livingstone, 'Submission to the Standing Committee on Foreign Affairs, Defence and Trade', p. 578.
43 R.C. Prendergast, 30 May 1994, in ibid., p. 586.
44 Adrian D'Hagé, 21 June 1994, in ibid., p. 588.
45 Senate Standing Committee on Foreign Affairs, Defence and Trade, 'Sexual Harassment in the Australian Defence Force', August 1994, p. 285.
46 Australian Human Rights and Equal Opportunity Commission, *Report of the Human Rights and Equal Opportunity Commission on Australian Defence Force Policy on Homosexuality*, pp. 1–20.
47 Smith, 'The dynamics of social change and the Australian Defence Force', p. 545.
48 Vanessa McQuarrie, 'The defence forces: Was the ban really lifted?', *Capital Q*, 29 April 1994, p. 7.
49 Ruth Pollard, 'Army action: Lesbian charged with barracks hug', *Sydney Star Observer*, 29 May 1997, p. 1.
50 'Sexual harassment in the Australian Defence Force', p. 279.
51 McQuarrie, 'The defence forces', p. 8.
52 'Just one of many cases', *Capital Q*, 29 April 1994, p. 7.
53 'Unacceptable sexual behaviour?', *Capital Q*, 15 March 1996, p. 1.
54 Ashley Hall, 'Military tackles anti-gay prejudice', *Melbourne Star Observer*, 28 April 2000, p. 3.
55 'Anger over troops' gay take-off', *Mercury*, 7 August 2004, p. 4.
56 Major-General P.J. Abigail, 'Report on the Bougainville Mardi Gras', 17 August 2000, Defence FOI 57/15/16.
57 'Anger over troops' gay take-off', p. 4.
58 Deborah Lupton, *Moral Threats and Dangerous Desires: AIDS in the News Media* (London: Taylor & Francis, 1994), p. 123.
59 See for example Martyn Goddard, 'We're in the Army now?', *Outrage*, April 1992, p. 40.
60 John O'Neill, 'Forces won't screen recruits', *Sydney Morning Herald*, 25 March 1987, p. 3; Scott, 'RAAF sacks AIDS carrier', p. 10; 'New defence policy on testing', *Sydney Star Observer*, 5 February 1988, p. 1.
61 DI(G) Pers 16–16, 'ADF Policy for the Detection, Prevention and Administrative Management of Human Immunodeficiency Virus (HIV) Infection', 30 March 1988, NAA A6721, 1985/18156 PART 2.
62 Australian Senate, 'Estimates Committee B: Report to the Senate', November 1992 answer to 23 September 1992 question on notice.
63 '36 personnel test positive', *At Ease*, July–September 1993, p. 7.
64 Tony Wright, 'Gay soldiers an AIDS risk: RSL chief', *Sydney Morning Herald*, 24 June 1994, p. 5.
65 'Call for defence force review', *Capital Q*, 21 October 1994, p. 3.
66 'Defence force in dock', *Sydney Star Observer*, 6 October 1994, p. 4.
67 Keith Scott, 'Defence force may face legal challenge over HIV applicants policy', *Canberra Times*, 5 October 1994, p. 3. See also Adrian D'Hagé, telephone interview with Noah Riseman, 24 April 2015.

68 Cheryl Ware, *HIV Survivors in Sydney: Memories of the Epidemic*, (Basingstoke: Palgrave Macmillan, , 2019), pp. 101–2.
69 W.J. Carter QC, HREOC Unlawful Dismissal Case, 1995.
70 'Prejudice on parade', *Campaign*, January 1996, p. 6. See also Dominic O'Grady, 'HIV ban "morally bankrupt"', *Sydney Star Observer*, 15 December 1995, p. 5.
71 *Commonwealth of Australia v. Human Rights and Equal Opportunity Commission and Another* [70 FCR 76] (1996).
72 Ruth Pollard, 'Federal Court backs soldier's dismissal, "unreasonable risk" claims Defence Force', *Sydney Star Observer*, 15 January 1998, p. 3; Gavin McGuren, 'Federal court rules against positive soldier', *Melbourne Star Observer*, 23 January 1998, p. 6.
73 *X v. The Commonwealth of Australia and Another* [200 CLR 177] (1999).
74 Alison Young, 'Into the blue: The image written on law', in *Cultural Analysis, Cultural Studies, and the Law: Moving Beyond Legal Realism*, ed. Austin Sarat and Jonathan Simon (Durham, NC, and London: Duke University Press, 2003), p. 336.
75 'Nelson', interview with Noah Riseman, 19 December 2018.
76 'Nelson', interview.
77 INDMAN Admin 1501A, 'Recognition of a person as family', 1986; Bruce Jones, 'In step with gay rights', *Sun Herald* (Sydney), 16 May 1993, p. 35.
78 David Anderson, *The Challenge of Military Service: Defence Personnel's Conditions in a Changing Social Context*, Background Paper No. 5, 1997–98 (Canberra: Australian Parliamentary Library, 1997); Ngaere B. Stewart, '"Married" to the military: The experiences of partners of Australian Defence Force members', BA Hons thesis, Edith Cowan University, 2004.
79 See for example, Donna Bull, 'Mobility: What effect does it have on ADF personnel?', *Australian Defence Force Journal*, no. 138 (1999), pp. 31–7; Edward Eastman, Robert P. Archer and J.D. Ball, 'Psychosocial and life stress characteristics of Navy families: Family Environment Scale and Life Experiences Scale findings', *Military Psychology* 2, no. 2 (1990): 113–27.
80 Keith Jeffers, interview.
81 Sherele Moody, 'Queer soldiers benefit push', *Sydney Star Observer*, 20 March 2003, p. 3.
82 'United Nations Human Rights Committee (UNHRC): Views on Edward Young v Australia', *International Legal Materials* 42, no. 5 (September 2003): 1244–57; Harley Dennett, 'Anzac's "gay desecration"', *Sydney Star Observer*, 26 April 2007, p. 1; 'Gay man denied war pension', *Melbourne Star Observer*, 5 May 2000, p. 3.
83 Edward Young, 'Veteran rights', *Melbourne Star Observer*, 18 August 2000, p. 4.
84 David Mitchell, interview with Noah Riseman, 10 July 2014, Brisbane. See also Riseman, Robinson and Willett, *Serving in Silence?*, pp. 188–90.
85 Group Captain A.J. White, 8 November 1993, courtesy David Mitchell.
86 *Australian*, 13 March 1995, p. 3.
87 Mike Seah, interview with Noah Riseman, 13 November 2016, Canberra.
88 David Mitchell, interview.
89 'Gays set up own defence support body', *Army*, 1 June 1994, p. 14.
90 'ADF members form gay support group: Press release', courtesy David Mitchell.
91 G-Force, 'Constitution of G-Force', 1994, courtesy David Mitchell.
92 'Support group for gay issues', *Navy News*, 2 June 1995, p. 11.
93 Karl Bryant, interview.
94 Peter O'Shea, 'A force within a force', *Capital Q*, 19 January 1996, p. 4; David Hannam, 'Fighting the defence forces', *Melbourne Star Observer*, 4 August 1995, p. 9.
95 Salter, *1995 Soldier Attitude and Opinion Survey: Equal Employment Opportunity Issues*, p. 9; *1995 Officer Attitude and Opinion Survey: Equal Employment Opportunity Issues*, p. 9.

96 Captain S.S. Pettersson, to Sergeant Karl Bryant, 15 March 1995, courtesy Karl Bryant.
97 Air Marshal L.B. Fisher, Chief of the Air Staff, to Sergeant K.W. Bryant, 26 April 1996, courtesy Karl Bryant. Mike Seah received almost the same response on 11 March 1996, courtesy Mike Seah.
98 General J.S. Baker, Chief of Defence Force, to Flight Lieutenant M.T.U. Seah, 'Application for redress of grievance', 18 October 1996, courtesy Mike Seah.
99 Jim Neely, Principal Solicitor for the Australian Government Solicitor, to Chris Sidoti, Human Rights Commissioner, 26 July 1999, courtesy Karl Bryant.
100 All of these documents courtesy Mike Seah; Mike Seah, interview.
101 David Mitchell, interview.
102 Senator Jocelyn Newman, Australia, Senate Questions on Notice, 'Australian Defence Force: Same sex next of kin', Question 1678, 10 May 2000, p. 14356.
103 David Mitchell to Brigadier Adrian D'Hagé, 30 August 1995, courtesy David Mitchell.
104 David Mitchell, interview. Mitchell's full story about the Mardi Gras is also available in Riseman, Robinson and Willett, *Serving in Silence?*, pp. 193–5.
105 Mike Seah, interview.
106 'Defence Force Mardi Gras float inappropriate: Chiefs', *Canberra Times*, 9 November 1995, p. 3.
107 Adrian D'Hagé, interview.
108 David Mitchell, interview.
109 Adrian D'Hagé, interview.
110 'G-Force join the parade', *PanDA* (Canberra), January 1996, p. 1.
111 Peter O'Shea, 'World first: Defence forces approve Mardi Gras float', *Capital Q*, 19 January 1996, p. 1. See also Peter O'Shea, 'Float go-ahead: "Yes, Private, you will go to the ball"', *Sydney Star Observer*, 19 January 1996, pp. 1, 3.
112 Christopher Zinn, 'Armed forces' gay float under fire from old soldiers', *Guardian*, 30 January 1996, p. 10.
113 Noah Riseman, '"Just another start to the denigration of Anzac Day"', p. 46.
114 NAA C475, 1718438.
115 David Mitchell, interview.
116 Mike Seah, interview.
117 Karl Bryant, interview.
118 Stuart O'Brien, interview with Noah Riseman, 25 August 2018, Canberra.
119 John Blaxland (ed.), *East Timor Intervention: A Retrospective on INTERFET* (Carlton, Vic: Melbourne University Press, 2015).
120 Keith Jeffers, interview.
121 Ron Bell, 'Democrats: Gay soldiers denied rights', *Melbourne Star Observer*, 15 October 1999, p. 4.
122 Moody, 'Queer soldier benefits push', p. 3.
123 Stuart O'Brien, interview.
124 DI(G) PERS 53–1: Recognition of Interdependent Partnerships, 1 December 2005.
125 'Gay couples set for defence rights win', *Sydney Morning Herald*, 22 October 2005, https://www.smh.com.au/national/gay-couples-set-for-defence-rights-win-20051022-gdmaou.html, retrieved 12 January 2020; Ian Gould, 'Defence rights win', *Sydney Star Observer*, 27 October 2005, p. 1.
126 'Fed: RSL supports equal treatment for same-sex partners in ADF', *AAP General News Wire*, 24 October 2005, p. 1.
127 'Gay couples set for defence rights win'; Gould, 'Defence rights win', p. 1.

Chapter 6

Transgender and intersex service

In July 2013 Army Lieutenant Colonel (later RAAF Group Captain) Catherine McGregor gave an eighteen-minute interview on the ABC television program *One Plus One*. McGregor was speechwriter to the Chief of Army, Lieutenant General David Morrison, and had penned his famous video address advising soldiers who could not respect women to 'get out' of the Australian Army.[1] McGregor candidly discussed her transition journey as a transgender woman serving as an officer in the Australian Army. She described receiving significant support and affirmation from Morrison and other members of the ADF, while also being the target of online abuse from other Defence personnel and civilians.[2] McGregor was subsequently profiled in numerous articles in the mainstream and LGBTI press,[3] appeared on the ABC television program *Q&A*, delivered an address at the National Press Club, had a column in the Fairfax press, regularly appeared in media as a political and cricket commentator, was the subject of a 2014 episode of *Australian Story*, was Queensland's nominee for the 2016 Australian of the Year (an honour that she later rescinded) and was the subject of the 2018 Sydney Theatre Company play, *Still Point Turning: The Catherine McGregor Story*. Although at times McGregor has expressed conservative views that have riled some members of Australia's LGBTI community, her visibility has unquestionably increased Australians' awareness of transgender people.

McGregor made an interesting point in her *One Plus One* interview; although she had been diagnosed with gender incongruence in 1985, she was not able to transition:

> The medical profession were not very empathic towards heterosexual males who presented with transgender feelings. They had a very stereotypical view of what a transsexual looked like in that era [1980s]. So if you were overtly gay, somewhat effeminate in presentation, and perhaps working in showgirl work as a transgender showgirl, or working on the fringes of the sex industry, they would transition you without quibbling. If you presented as I did, as someone who was functioning as a heterosexual male, they were very reluctant to recommend it.[4]

Oral histories with older transgender women and clinical understandings of what was then called 'gender identity disorder' broadly support McGregor's statement, but digging deeper into the history of transgender health reveals a different story. As early as 1988, American psychiatrist George R. Brown published research which found that transgender women disproportionately served in the US armed forces. Military service represented what he called a 'flight to hypermasculinity': in trying to deny their authentic female selves, pre-transition transgender women threw themselves into hyper-masculine environments like the armed forces, trying to 'prove' their masculinity.[5] Later research by Brown and others found that many transgender service members, unable and/or unwilling to transition, had suicidal ideation and would serve in combat roles, often volunteering for dangerous missions.[6] Research from other countries, including Australia, supports Brown's conclusions about the historical attraction of defence forces for people questioning their gender identities. Transgender men, too, have often enlisted because militaries are spaces where women can more safely express masculine behaviour.[7] The logical extrapolation from such research is that there is a long history of transgender people serving in the ADF.

This chapter explores the hidden history of transgender military service in the Australian armed forces, as well as the even more obscured history of intersex service. It begins with the fragmented histories of transgender service dating to the immediate post-war period, featured in occasional newspaper reports and in military police investigations or contained within transgender oral histories. The ADF did not actively think about transgender people until the 1990s, the first formal policy ban being implemented in 2000. There has never been a formal policy on intersex service. Because intersex variations have been treated as a medical issue—rather than as one of sexuality or gender diversity—their experiences have been rendered even more invisible. Even so, intersex service experiences still highlight the very rigid understandings of sex, gender and sexuality that have, for so much of the Australian military's history, rendered the institution intolerant of diverse bodies, genders and sexualities.

A note on terminology

Terminology is always challenging when discussing issues around gender identity and intersex variations. Language has evolved rapidly, with some terms that were previously acceptable (e.g. transsexual) now considered inappropriate. Complicating the issue further is that some transgender or intersex people might still use older terms to describe themselves, and it would not be appropriate to challenge someone's self-identity. When looking at examples from the past, it can be problematic to apply present-day concepts retrospectively. The term 'transgender' was not coined until 1965 by psychiatrist John Oliven to describe those males who dressed and lived as women but did not desire gender affirmation surgery. This was in contrast to those who did have gender affirmation surgery (transsexuals) and those who enjoyed dressing as women but did not wish to live full-time as women (transvestites).[8] These terms almost always described trans women; only in the 1980s did the media and scholarly literature begin to mention trans men (and rarely at that), and the idea of being non-binary has only emerged slowly since the 1990s.[9] In the 1990s the word 'transgender' became a popular umbrella term to encompass all people whose gender expression or identity differed from their sex assigned at birth, regardless of surgical or medical interventions. 'Cisgender' is the term used to describe those people who are not transgender, meaning that their gender identity aligns with their sex assigned at birth.

When looking at past examples, we must be cautious not to impose labels on individuals when there is no indication of how they saw themselves. Where we do find examples or references to outdated terms like 'transvestite' or 'transsexual', however, we are able to examine them within the remit of historian Susan Stryker's umbrella definition of transgender—'people who cross over (*trans-*) the boundaries constructed by their culture to define and contain that gender':

> Some people move away from their birth-assigned gender because they feel strongly that they properly belong to another gender through which it would be better for them to live; others want to strike out toward some new location, some space not yet clearly described or concretely occupied; still others simply feel the need to challenge the conventional expectations bound up with the gender that was initially put upon them. In any case, it is the movement across a socially imposed boundary away from an unchosen starting place, rather than any particular destination or mode of transition.[10]

We are conscious of adopting the 'GLAAD Media Reference Guide—Transgender', which describes best practice terminology. We also avoid terms like 'cross-dresser' and instead use 'dresser'.[11] However, when an oral history or historical document uses outdated terminology, we preserve the original language for the sake of historical accuracy and to provide context of the era. It is generally considered disrespectful to use transgender people's pre-transition names, sometimes referred to as 'deadnames'. We have tried to follow this practice as much as possible. In a handful of cases, however, we reference the pre-transition name when the person is deceased, their previous name is significant, or media reports mentioned only the pre-transition name. We also use pronouns that align with Defence members' affirmed gender, even when referring to their pre-transition selves. When their gender identity is ambiguous or unknown, we use the non-gendered pronouns they/them.

Transgender and intersex are not the same. The definition of intersex provided by Intersex Human Rights Australia (formerly Organisation Intersex Australia) is:

> Intersex people are born with physical sex characteristics that don't fit medical and social norms for female or male bodies. It is worth taking time to recognise that this definition does not specify a sex, legal sex classification, birth sex assignment, gender, gender identity or sexual orientation. Intersex people can be heterosexual or not, and cisgender (identify with sex assigned at birth) or not.[12]

Although some intersex people may transition genders and therefore *also* identify as transgender—and these were the intersex people more likely to attract the attention of ADF authorities or the media—the majority of intersex people are cisgender. Intersex activists are rightly concerned that transgender and intersex not be conflated. Intersex variations are about bodily diversity while the term transgender refers to gender diversity. Yet intersex and transgender people have an important commonality, which is why we have included them in the same chapter: both transgender and intersex people challenge long-held ideas about gender, sex and the body. Such challenges are especially problematic for a highly binary-gendered institution such as the ADF. How the ADF has grappled with these challenges and how transgender and intersex members have navigated the institution's policies is at the heart of this chapter.

Fragments of transgender service

German sexologist Magnus Hirschfeld first coined the term 'transvestite' in 1910 to refer to men who felt an internal urge to dress in women's clothing.[13] In the 1930s European doctors performed the first gender affirmation operations, but they did not become widely known and more readily available until after the operation on American World War II veteran Christine Jorgensen in 1953. Harry Benjamin popularised the term 'transsexual' in 1954, continuing the medicalisation of transgender people, which activists still challenge.[14] While sexologists, psychiatrists and psychologists were distinguishing transvestites/transsexuals from homosexuals, popular understandings tended to conflate them. In other words, Australian society tended to view transvestites and transsexuals as homosexuals who had simply gone further in their effeminate demeanour. Because of this popular association of transgender with homosexuality, the Australian services never developed specific policies on transgender the same way as they did for homosexuality. As the following cases show, transgender people would either be treated the same as suspected gays and lesbians or, in some cases, were dealt with medically.

Globally, tales of women impersonating men to go to war date back to ancient times. It is only recently that historians have begun to question these subjects' gender identities, never definitively labelling them but rather imagining trans possibilities. Perhaps the earliest Australian example was Maud Butler, who was caught attempting to enlist in World War I as a male.[15] An anecdote published in the newsletter of the Chameleons Society of Western Australia described one member who served in the Australian Army in World War II and was seconded to a British regiment on the Rhine:

> Well, one day during a routine inspection, the RSM was going through the member's kits and came across a whole heap of ladies apparel. The RSM grunted and said 'You are Australian, aren't you?' 'Yes, Sir,' said our member. The RSM grunted and moved on, leaving our member wondering what in the hell was going to happen. Well, what did happen was rather remarkable. A few days later the RSM came across our member and pressed a card into his hand and said, 'I can recommend this club.' On it was a[n] address of a crossdressing club in Hamburg.[16]

In World War II, another site of trans possibilities was drag performances. Drag queens/kings are people who perform the persona and dress of another gender, but they do not identify with that gender the way transgender people do. That said, historically, many transgender people found drag to be an entry point to experiment with gender expression. As Yorick Smaal notes, drag performances allowed servicemen 'to enjoy the accoutrements of queer life—powder, paint and lipstick—as well as hobbies like sewing—without fear of official recourse'.[17]

More definitive examples of transgender service members are from the post-war era when publicity around Christine Jorgensen and the category 'transsexual' gave transgender people new language. In August–September 1955, *Truth* newspapers ran the headline 'Ex-RAAF sergeant will become a woman'. The articles told the story of Robert Brooks and their lifetime struggle with gender and longstanding urge to dress in women's clothes. Brooks purportedly 'despised effeminate men' and constantly aimed to be located in masculine environments. It is for this reason that they enlisted in the RAAF:

> I thought there might be something about my home life which was making me that way … So I joined the RAAF in 1949 and served for six years and four months, retiring with the rank of sergeant major. But still there was this conflict inside me. I kept to myself, I had no friends, and I lived a solitary life.

A psychiatrist diagnosed Brooks as a transsexual, and Brooks left the RAAF and intended to travel to Denmark for gender affirmation surgery. Reportedly, Brooks intended to take the name Christine as an homage to Christine Jorgensen.[18]

Over the years, especially from the mid-1970s on, newspaper articles about transgender women or publications by transgender organisations would make passing reference to a person who had served in the armed forces. One of the first members of Sydney's Seahorse Club used to regale members with stories of being ejected from a plane during the Korean War and surviving the hazards of the China Sea.[19] In 1976 Perth newspaper *The Independent* ran a feature about a transgender organisation, the Chameleon Society of Western Australia. The article noted the diverse career backgrounds of the Chameleon Society's members, including 'a retired Royal Navy submarine commander, a chief surveying engineer, RAF psychiatric nurse, [and] a former lieutenant colonel'.[20] Another British immigrant to Australia, Susan Williams, wrote in the Seahorse NSW magazine *Feminique*: 'Between being in the army and coming to Australia in 1959 I wore female underwear most times, more so in the winter months when outer clothing was more bulky.'[21] A Seahorse Victoria profile of member Lynette Jones mentioned that 'Lynette saw 4½ years of Army service, was in Japan with the occupation forces, and worked with American counter-intelligence on the Japanese war crimes trials'.[22] Chameleon Society member Robyn served in the Navy in the 1950s, including as crew when HMAS *Fremantle* was sent to clean up British atomic bomb tests on the Monte Bello Islands.[23] Marie-Desirée D'Orsay-Lawrence, secretary of the South Australian-based transgender group Chameleons, briefly alluded to her Army service jokingly when she appeared before a hearing of the Senate Inquiry into Sexuality Discrimination in 1996:

> You are transgendered if you dress in the apparel of the opposite sex. Whether you are a transvestite, a cross-dresser or transsexual, or just like wearing pantyhose. Quite a number of the members of the army must be transgendered because a lot of them wear pantyhose on winter mornings out in the field. I know because I used to give them to them. I was in the medical corps.[24]

The haphazard way in which these narratives mention military service suggests that while it was part of their life journeys, it was not necessarily central to their identities as transgender people.

Other first-hand accounts linked military service to a wider attempt to repress femininity, aligning with Brown's concept of the flight to hypermasculinity. Kaye Rona Ellis's narrative, published in Seahorse Victoria's magazine *Seahorse Scene*, described how she 'tried right up to the very end to live the part of a "Pseudo Male", knowing that I was attempting to "murder" the real me, just so that a foriegner [*sic*]—a man, whose body held me in captivity, may live'. Among the blokey professions Ellis held, she mentioned being a soldier (as well as fireman, stoker, truck driver and shearer).[25] Another transgender veteran, Robyn, used to dress in her sister's clothes as a child, which drew the ire of her father. Robyn joined the Army and even served in Vietnam, saying in an interview with the *Sunday Age*: 'I thought if I joined the Army, I would prove to my father that I was a man. I was trying hard to present an image that wasn't me. I was accepted by other men, but I wasn't one of them.'[26] A newspaper article about Beth noted that she, too, had a father who pressured her into rejecting feminine interests and presenting a chauvinist persona. Beth joined the RAAF when she turned eighteen because: 'That macho environment was another way to hide myself. There's no way you'd expect a transsexual to be in a military type of environment.'[27] Gina Mather, president of the Australian Transgender Support Association of Queensland, testified before the 1996 Senate Committee Inquiry into Sexuality Discrimination:

> I could not understand why I was a bit different to other people. I never got married, even though I had opportunities to get married. I played footy for Queensland. I did all the macho things. I went overseas. I spent 19 months in Vietnam and yet I could not

> get rid of this underlying desire to be a member of the opposite sex.[28]

Other transgender people were compelled to enter the armed forces as national servicemen. Almost all Australian males aged eighteen were compelled to undertake 176 days training as national servicemen from 1951 to 1959. From 1965 to 1972, under a new National Service scheme, all twenty-year-old males had to register for a birthday ballot and could be called up for military service. Katherine Cummings was among those who served in the first incarnation of National Service from 1954 to 1956 in the RAN. Cummings writes positively about her time as a sailor, when she learned important skills such as typing and became quite fit. The most memorable part of Cummings's service was on HMAS *Australia* as escort cruiser for part of Queen Elizabeth and Prince Philip's royal visit in 1954. Cummings was already dressing in women's clothes before joining the RAN, but found it difficult to do so during her service. She feared being caught but it was also difficult logistically. She writes,

> Naval lockers are minuscule and completely filled by one's uniforms and equipment, as well as being liable to inspection at any time, so that there was no way I could keep any feminine items by me. Nor was there a moment's privacy, as about twenty of us shared a dormitory where we lived, slept, ironed and mended when we were not on duty.[29]

That said, Cummings notes that on rare occasions during leave or while attending university, she would dress in secret and take photographs.

'Penelope' was called up in one of the first intakes of the new National Service scheme in 1965. Penelope's enlistment coincided with a period in her life when she was trying to repress her female identity, so she became heavily involved in playing Australian Rules Football and resisted the temptation to dress in women's clothes. Penelope married just before enlistment because of an unplanned pregnancy. The marriage, according to Penelope, worked in her favour; she was originally slated to be posted to Vietnam with the 6th Battalion, Royal Australian Regiment, but at the last minute she was instead posted as a clerk at an office in Melbourne—supposedly because authorities did not want to send a married national serviceman in the first deployment to Vietnam.[30] For Penelope's two years of National Service, she mostly worked with members of the WRAAC. This helped alleviate her gender distress:

> Mixing with the WRAACs and talking with them on more than just a casual basis, if you know what I mean, made me feel like I

> was one of them or I was accepted by one of them and that was sort of a trade-off, if that's a way to explain it. That I—'Oh yeah, I might look like this but underneath I'm one of you guys.' Never told them it and they didn't know but it made me feel better.[31]

Penelope continued to suppress her urges to dress and would only resume the practice years after she discharged.

Just as being openly gay could have someone rejected from National Service, so too could being transgender. A Draft Resisters Union flyer titled 'How not to join the Army' suggested:

> BE GAY: Play the homosexual bit ... Wear white slacks, have your hair cut rather camp, wear a charm. Visit a couple of camp pubs and study homosexuals. Learn the gestures, the wrist movements. And the delicate body movements, how to touch the fellow you're talking to suggestively, how to smoke a cigarette. Be a little pathetic, talk melodically, act embarrassed in front of the other inductees when you undress. Ask your girlfriend to show how.[32]

Openly transgender people could go one step further: arrive at their National Service medical examinations dressed as women. Ronnie Ahlburg, who starred in the cabaret revue Les Girls in St Kilda, announced to the newspaper *Truth* that they intended to show up in 'a new, glamourous outfit designed for the occasion. It'll be stunning. And it'll impress the doctors, too.'[33] Ahlburg attended the medical examination in a green miniskirt, snake-skin boots and midi coat and carrying a matching handbag. The doctors were reportedly kind to Ahlburg, although the medical check was embarrassing. Ahlburg was quoted as saying: 'I did an aptitude test, which said that if they took me I could only be an entertainer ... I know I couldn't do anything rugged—I'm not strong enough or manly enough.'[34] Ahlburg's fate is unknown, but their story represents an important note about how savvy, brave transgender people could use the media and their gender identity to their advantage in certain contexts.

Hiding and outing

Transgender people were never in the sights of the service police and would likely have been considered just to be homosexuals. Still, some service police investigations specifically mention 'transvestites'. A December 1971 investigation into the theft of numerous bras, panties and other undergarments from a clothes line at the WRAAF barracks at RAAF Base Pearce concluded 'that an airman or airmen suffering from a mental aberration is responsible for the thefts'.[35] Similar unsolved thefts of WRAAF members' clothing, undergarments and jewellery happened at RAAF Base Penrith during the period 1970–72.[36] In November 1972, service police arrested a leading aircraftman who was caught in the WRAAF barracks in women's clothing. The culprit confessed to stealing female clothing for more than twelve months, and most of the clothing was recovered in their room. The commanding officer referred the leading aircraftman for psychiatric evaluation and intended to assess their employment suitability through an administrative rather than a disciplinary process.[37] This early example represents a similar approach as to gay servicemen during the same era.

During the era of the heightened LGB ban, transgender people could be caught in service police investigations because they were presumed to be homosexuals. Nikki Bobbermien knew from an early age that she was feminine, and from age fourteen she also knew that she was attracted to males. She joined the Army as an apprentice chef because she wanted to avoid what she saw as the trap common to so many gay men and transgender women who lived on the margins of society: sex work and drug addiction. While serving at Oakey Army Aviation base outside Toowoomba, Bobbermien would regularly drive to Brisbane at night and perform as a female at the Empire Bar. One night the RAAF Police followed Bobbermien and arrested her after a show. Bobbermien survived a suicide attempt and then commenced hormone therapy shortly before her discharge.[38] Details about Bobbermien's police case and discharge are unavailable, but the surveillance and rough treatment by the RAAF Police suggest that she likely went through a similar process to LGB service members as outlined in chapter 3.

Hiding and secrecy were, of course, essential for transgender people in the armed forces. Just as Katherine Cummings would sometimes find opportunities to dress away from base, so, too, did Beth buy women's clothes while serving in the RAAF from 1981 to 1991: 'When the "fits of maleness" were getting me down, dressing up made me relax but it didn't fulfil me. It was basically a façade where I could see myself as partially presentable, [yet] I still knew things weren't quite right.'[39] 'Kate' served as a midshipman from 1983 to 1985 and recalls a horrific time being bullied and abused during officer training at HMAS Creswell. Very early on, Kate learned how homophobic and transphobic the Navy was: 'So, you were then taught how to teach your sailors as a DO [divisional officer], how to recognise the transgender person. In fact, the Navy had an official name for it: they called transgender Benny Boys. They'd been a boy and now is a girl.'[40]

Although Kate never hinted at her transgender identity, early on her abusers accused her of being female. Unlike some of the above examples of flights to hypermasculinity, Kate was the opposite: she eschewed masculine stereotypes, including playing sports like Rugby, and did not drink alcohol. Kate would only dress in private when she had opportunities off base. She explains the feeling of self-hatred when she dressed:

> But you did it, and the worst part was you didn't know why you did it, and that made it even more worse. So you'd do it, you'd feel guilty, you'd ... feel revulsed [*sic*], and you'd throw the clothes away. Then a little while later you'd do it all over again. And you're left wondering: why am I behaving the way I am?[41]

Kate requested her own discharge after two and a half years, but did not transition until years later.

Some transgender people came out to secure their discharge. Janette Ridge enlisted in the Army in 1975 'to prove something to myself ... I think I have known since the age of eight that there was something different about me—and the Army just confirmed it.' Ridge suffered insults from other soldiers for two years before she requested a discharge and then began transitioning.[42] Allan Royce Jaynes married and joined the RAAF in an effort to repress her transgender self. Yet the gender distress did not go away and affected her work. She staged her own disappearance from Windsor, NSW, but was picked up by police while working in Queensland. Jaynes came out as transgender and secured a discharge at her own request before commencing her transition to 'Vicki'.[43]

Just as being an officer could provide a veneer of protection for LGB people so, too, did transgender officers have slightly favourable treatment. Alex Pollock enlisted in the RAAF in 1960. The psychologist's report for prospective air crews described Pollock as 'a rather easy-going, quietly spoken, personable young man who generally establishes good social relationships with others, with the exception of his step-father'. Pollock served as a navigator, including training in the United States on F-111s in 1967–68. From 1961 to 1977, Pollock consistently received praise in annual evaluation reports, including several commenting on her social wit and demeanour. Pollock was even made a member of the Order of Australia—Military Division in 1976 and was promoted to the rank of wing commander later that year.[44]

In April 1977, Pollock's marriage broke down. According to her friend Catherine Heal, Pollock came out as transgender to her wife. The wife left Pollock and took the children overseas, threatening to out Pollock if she contested this. Pollock attempted suicide and survived only because a RAAF friend found her.[45] At some stage in 1977, Pollock came out to her superiors as transgender. Rather than being discharged, however, they permitted her to remain in the RAAF until she attained twenty years' service. Air Vice Marshal F.W. Barnes wrote on Pollock's 1978 annual evaluation report:

> … he is being allowed to continue to serve in the interests of his family and on the understanding that he maintains an impeccable façade … I agree that Pollock continues to act out the life of an operationally oriented, gregarious and dedicated officer in a most skilled and convincing fashion; to the extent that he effectively and efficiently operates his unit. However, I cannot accept the Part 3 report [comments earlier in the evaluation] and I will watch Pollock's performance this year with some interest.

Pollock discharged on 31 January 1980 and noted in a letter accompanying her exit survey sheet: 'My main reason for resigning was very personal and had nothing to do with the various shortcomings—real or imaginary—of the RAAF.'[46] Pollock's treatment is interesting because superiors knew about and even openly disapproved of her transgender identity, but still allowed her to serve (albeit under a close eye). It is not clear whether this was because of her high rank, her extensive years of service, personal relationships with the RAAF hierarchy, sympathy for her difficult circumstances or some combination of the above. Pollock transitioned later in the 1980s and severed nearly all ties to her RAAF life, keeping in touch with only two mates. She told very few people about her past RAAF life, but revealed the story to her friend and confidant Catherine Heal a few years before she passed away in June 2005.[47]

Most outed transgender service members were not as lucky as Pollock. 'Helen' joined the Navy in 1976 at age seventeen and trained as a radar operator. Helen did several tours on the naval flagship aircraft carrier HMAS *Melbourne*, but also spent much time based at Watsons Bay in Sydney. She would venture into Kings Cross bars like the Carousel and see performances like Les Girls, exposing her to trans women. To mask her own gender distress, Helen turned to alcohol, and by 1979 she was experiencing significant internal conflict. Helen came across a very small transgender activist group in Sydney and attended a few of their meetings. Finally, in 1980 Helen confessed to a medical officer that she was transgender. The doctor's reaction was wholly negative; the doctor transferred Helen to the hospital at HMAS Penguin. Helen continued to work at that base while undergoing numerous psychological and psychiatric examinations over several months. As she explains, 'So, their approach to me was that I was being dishonest [to secure an early discharge] … There was no care in it about me. It was all about am I being fraudulent.' Once the psychologists and psychiatrists were convinced that Helen was genuinely transgender, they discharged her medically in November 1980. Helen felt significant stigma attached to her discharge, to the point that for years she would not join the RSL or other service organisations.[48]

One theme that permeates all of these examples is the inconsistency of approaches to transgender service members. Whether treated medically, discharged swiftly at their own request, following the patterns of LGB personnel or allowed to continue serving, the fact that there was no clear policy or direction meant that there were a variety of experiences, treatments and potential outcomes. Indeed, the practices around transgender service members from World War II through the 1990s was similar to gay and bisexual men before 1974: discretion could mean a blind eye turned and scope for commanding officers to address each case differently. It would be only after transgender visibility, activism and rights slowly emerged in Australia in the 1990s that the ADF would at last adopt formal policies and procedures on transgender service.

Intersex experiences

Intersex awareness has always been even more on the margins than transgender visibility. Partly this is due to the small percentage of the population who have intersex variations, estimated at 1.7 per cent by Intersex Human Rights Australia. There are at least forty different intersex variations, which can all manifest in different ways.[49] Some of these variations may be identifiable at birth, which has led to the controversial practice of surgery on intersex infants (banning non consensual surgery on intersex children and infants is at the core of intersex activism, as outlined in the Darlington Statement[50]).

Just the language around intersex is very recent and evolving. Although geneticist Richard Goldschmidt first used the term in 1917 to describe people previously referred to as 'hermaphrodites', it was not until after World War II that the medical profession began to adopt the terminology.[51] In popular discourse, 'hermaphroditism' continued to be the dominant language even though this terminology is generally considered offensive by people with intersex variations. Australia and the world's first intersex group was the Androgen Insensitivity Syndrome Support Group Australia, founded in 1985, while the first known organisation to use the term 'intersex' was the Intersex Society of North America, founded in 1993. In Australia the 'I' for intersex began to be incorporated in the rainbow alphabet in the early 2000s, and it is then that the word 'intersex' became more common in public discourse.

Intersex people have been invisible for other reasons beyond just the small numbers and ignorance about intersex variations. Other reasons include: the stigma attached to intersex variations; secrecy surrounding intersex variations and surgery performed on infants; and the diversity of variations that only in the last thirty years came to be grouped under the umbrella of intersex. Some intersex people go for years or even their whole lives unaware of their intersex variations. That was certainly the case for one of the few intersex ex-service members identified in this project: Peter Stirling. He served in the WRAAF from 1954 until 1955 as a female, unaware that he was living with an intersex variation. Stirling was actually discharged from the RAAF for having a relationship with a woman, meaning that his time in the ADF aligned more with the lesbian and bisexual servicewomen's experiences outlined in chapter 2. About a decade after his discharge, as Stirling's body continued to change, he was diagnosed with Klinefelter syndrome (having XXY chromosomes) and was given hormone and surgical treatment in London to transition to male.[52] Stirling's full life has thus traversed the categories of lesbian, bisexual, transgender and intersex.

Stirling's story is in many ways reflective of the broader historical patterns around intersex people's service. Intersex variations are biological and embodied, whereas sexuality and gender identity are internal and social–psychological. Sexuality and gender identity are expressed through behaviour (e.g. same-sex encounters; dressing), which is why the ADF was able to formulate regulations to police and ban the behaviour (or, more accurately, the people who performed the behaviour). Intersex variations were medical issues. In most instances an intersex person who was physically fit could pass the medical examinations and had no trouble enlisting in the ADF. They would be required to identify within the gender binaries of the ADF (male/female), but this was normally not a problem. One myth about intersex people is that they are all non-binary, but the majority identify as cisgender male or female. Indeed, 'Gina', our one intersex interview participant, recalls at least three other cisgender intersex people serving in the RAAF in the 1970s—one female and two males.[53]

Like Peter Stirling, Gina did not know that they were intersex when they first enlisted in the RAAF in 1971. Gina does not identify with a particular gender, but they served as a male. Gina entered the electronics mustering and worked as a ground tech, meaning working on the electronic equipment on the ground (as opposed to on aircraft). Gina was involved in the RAAF relief effort for Cyclone Tracy in 1974–75, assisting in the transport of equipment and then installing the quad radar and other equipment at the relief base in Darwin. Gina had a troubled upbringing and was socially awkward, finding it difficult to make friends in the RAAF. Gina also did not know that they were intersex, as they never questioned why their body was different from others.

Gina says that service members sometimes ridiculed others for their physical appearance when they did not fit into sex stereotypes of the era. One of the two intersex men whom Gina encountered in the RAAF was the target of physical abuse during recruit training at RAAF Base Edinburgh because of his feminine appearance and very tiny genitals. Gina remembers that he had a derogatory nickname that followed him throughout his career, but he served his full six-year term in signals. Gina recollects that the other intersex man ran into difficulties when he began to exhibit female secondary sex characteristics: 'This chap was very, very feminine and started developing breasts, and the breasts got bigger and bigger and more and more obvious. And [he] was held back a course and then held back another course and was forever on medical parade and then just quietly disappeared off the course under me, never to be seen again.'[54]

Gina was adept at avoiding situations where other service members could see them nude. When questioned about communal showers, particularly during recruits training at RAAF Base Edinburgh, Gina explained: 'I actually had a strategy around that—which was getting in early or getting in late. They had little barriers between each one and all that sort of stuff. I was fairly deft at wending my way around that. It only lasted for ten weeks. After that, it was all private showers with doors on it.' Gina's body only once attracted the attention of other airmen, and this set in motion the events leading to their discharge. During one morning parade the warrant officer asked:

> 'When was the last time you shaved, Airman?' I said, 'I've never shaved in my life.' [Laughs] So I was marched up with the flight sergeant—I was marched up to the ASCO [airmen's canteen] … where I was made to buy shaving cream, a razorblade, all that sort of stuff, marched into the bathroom back at base radio. He had to stand there and watch while I shaved: 'And you will do this every morning.'[55]

About three weeks later, Gina was summoned for what they describe as a 'fairly invasive' physical examination. The doctor referred Gina to an endocrinologist, who prescribed testosterone. The hormones did produce the facial hair but at the cost of some nasty side effects, including blinding headaches, bloating, anger and jitteriness. The doctors could not figure out what was wrong with Gina but said that they had 'glandular' problems and was unfit to continue serving. Gina was medically discharged in 1977, still not knowing what medical condition affected them. In 1998 doctors finally diagnosed Gina with androgen insensitivity syndrome, and even then the doctors did not tell Gina that this was an intersex variation. It was only when Gina returned home and searched on the internet that the word 'intersex' came up in several thousand hits. Gina would go on to be involved in intersex activism in the 2000s.[56]

The final example we have of intersex service is of Army soldier Tony Farrugia, who enlisted in 1991. Since puberty, Farrugia's body had undergone unusual changes: painful internal bleeding once a month, which would clot in the abdomen because it had nowhere to escape, and surgical removal of a breast during their teenage years. Still, Farrugia had a functioning penis and one testicle and identified as male, repressing both homosexual thoughts and an internal feeling of being female. Other recruits ostracised Farrugia during recruit training at Kapooka because they were not aggressive enough. Rumours circulated that Farrugia was both gay and a 'transvestite', and eventually they were kicked out of the barracks. At Farrugia's subsequent posting in Enoggera, again the other service members—both male and female—taunted them.[57] The ridicule that Farrugia attracted aligns with Gina's examples about how embodied ideas of gender and sexuality could affect intersex personnel who did not align with social expectations assigned to their bodies.

In June 1992 an ultrasound identified that Farrugia had a uterus, and doctors diagnosed that they were a 'hermaphrodite'. Farrugia was quoted as saying, 'I've always hated myself for what I thought I was, so you can imagine the relief I felt when I discovered why I had these feelings.'[58] Four soldiers who learned about Farrugia's intersex variation called them a freak and threatened to assault them, leading authorities to move Farrugia off base while they underwent further physical and psychological tests. Farrugia intended to transition to female and wanted to continue serving in the Army. They commented: 'I haven't done anything wrong because when I signed on the dotted line my birth certificate said I was a male. If they get rid of my career then I want to be compensated for it. I'm a pretty tough guy but I'm also a super-bitch.'[59] Farrugia discharged on 27 November 1992; the discharge certificate lists the reason as AMR 176(1)(H): 'that the soldier is medically unfit'.[60] Farrugia's fate after leaving the Army, including whether they transitioned gender, is unknown.

The examples of Peter Stirling, Tony Farrugia and 'Gina' collectively reveal the complex interweaving of medicine, social constructs of gender and ADF policies on homosexuality. Those intersex service members whose gender identities did not align with their sex assigned (or surgically imposed) at birth could be dealt with under the policies on homosexuality. Those who were cisgender and heterosexual might have served with uneventful careers, particularly if their intersex variations did not attract attention from colleagues or superior officers. Those intersex members whose variations were visible, on the other hand, could be targeted for derision or ridicule. Amid all of this, though, as Gina summarises, 'Nobody talked about intersex in those days, and anybody who said they did is a liar.'[61] As chapter 7 explains, it would not be until the 2010s that the word 'intersex' entered the ADF's lexicon alongside 'lesbian', 'gay', 'bisexual' and 'transgender'.

Banning transgender service

Although intersex Australians continued to be invisible in the 1990s, that was an era of rising transgender activism for anti-discrimination protections and legal recognition in their affirmed genders. The first time that the ADF formally addressed transgender military service was in a submission and testimony before the 1996 Senate Inquiry into Sexuality Discrimination. The inquiry into proposed legislation by the Australian Democrats would have banned discrimination against people on the grounds of their sexuality or gender identity.[62] The joint ADF and Department of Defence submission noted that the bill's main consequences for LGB members would be to recognise same-sex relationships and extend all related benefits to those couples. The crux of the submission, though, was to oppose the bill because of the ramifications for transgender service. The submission described the ADF's contemporaneous policy on transgender personnel as 'requir[ing] members to wear the uniform of their biological sex, until their gender reassignment has taken place ... In view of the operational and managerial issues involved, the Australian Defence Force strongly believes that recognition of transgender identity should only occur after gender reassignment has successfully taken place.'[63]

The ADF submission and Commodore Jim O'Hara's testimony before the Senate committee outlined three grounds why they opposed the bill:

1. Women might self-identify as men to apply for combat roles, and men might self-identify as women to avoid combat.
2. Transgender service members would adversely affect troop morale and group cohesion (the same arguments used for years against LGB service).
3. The definition of transgender was problematic because it did not require any medical assessments to confirm someone's gender identity.

O'Hara said that the ADF had no qualms with transgender people but rather with the bill's definition, which would permit people to nominate their gender.[64] The ADF submission, as well as concerns about transgender women in sport, became lightning rods that conservative politicians and pundits used to attack the Sexuality Discrimination Bill. Columnist Piers Akerman derided the notion of transgender people serving: 'A squad of cross-dressing SAS [Special Air Service Regiment] troops storming through the bush sounds like something out of Monty Python.'[65] Victorian RSL President Bruce Ruxton wrote, 'If the Senate wears this Bill it will wear anything. A man is a man and a woman is a woman. A man who dresses as a woman defies description.'[66] Although the final inquiry recommended passage of the Sexuality Discrimination Bill, neither the government nor the ALP opposition supported it and the bill never went ahead.

In April 2000 the ADF adopted its first formal policy on transgender service: Defence Instruction (General) Personnel 16–16 (hereafter DI(G) Pers 16–16), 'Trans-gender Personnel in the Australian Defence Force'. The document defined and pathologised transgender around medical interventions and gender affirmation surgery, which set an important precedent for later policies. Unlike other medical or psychological challenges confronting Defence members, though, the ADF would not support members who were living with gender incongruence. The policy was divided into a section around recruitment of transgender service members and a part about management of serving members who wished to transition. The recruitment section said: 'A person undergoing or contemplating gender reassignment cannot be considered suitable for service in the ADF because of the need for ongoing treatment and/or the presence of a psychiatric disorder.'[67]

The section about management of serving members said that all people who came out as transgender would be referred for medical assessments of their suitability to continue serving. Subsequently:

> Trans-gender members who are assessed as being medically unfit to continue serving or who do not meet the ADF individual readiness requirements may be discharged. Also, discharge will be considered where the member's behaviour in identifying as a member of the opposite gender, while remaining legally and anatomically a member of their biological gender, could have serious consequences upon operational effectiveness, cohesion, morale and discipline of any unit to which the person is posted.[68]

The policy implied that transgender people could continue to serve in some circumstances, similar to the way the 1974 policy on homosexuality said that 'unconfirmed homosexuals' could be retained in the forces. Just as that policy became a blanket ban on LGB members, so too did DI(G) Pers 16–16 essentially become a ban on transgender service. As such, those members who wished to transition had the choice either to hide their authentic gender identity or to come out and discharge.

DI(G) Pers 16–16 retained one element of the practices outlined in the 1996 submissions to the Inquiry into Sexuality Discrimination: 'A member who is discharged in the above circumstances and subsequently undertakes successful gender reassignment surgery, may apply to rejoin the ADF as a person of their new gender.'[69] 'Mitch' is one person who did this; he served in the Army Band from 2000 to 2003. In 2002 he began taking testosterone. Theoretically any full-time ADF member is meant to see Defence doctors in the first instance, but for this Mitch saw an outside specialist on the sly. Mitch covered up most of his bodily changes and explained away his changing voice as a consequence of talking too much. Uncomfortable with Army life, he left at the end of 2003, never outing himself as transgender. In 2005 Mitch decided to rejoin the Army Reserve Band as a male. Mitch's commanding officer was supportive and sent him to the base doctor. Mitch recalls the reaction: 'He said, "I hope you do realise that as part of this transition that even though you will be male you won't be eligible for the SAS." And I just looked at him dead straight and I said, "Oh, well, that's just fucked, isn't it, because that was my life ambition. I want to be a commando."' Mitch passed his fitness test, and as he had already updated his legal documents, the Army updated his name and gender with little hassle (although Mitch notes that someone had shoved a newspaper article about female-to-male transgender people in his medical file).

In 2009 Mitch switched back to full-time service without any problems and was posted as an instructor at the School of Music in Watsonia, Vic. Given the small field in which he worked, everyone knew who Mitch was.[70] Mitch's relatively easy re-enlistment is likely because he was in a non-combat, specialist role as a musician. Moreover, by starting as a reservist, the physical requirements were not the same as for a full-time or combat role. It is unknown whether a trans woman, non-specialist or full-time member would have had such a smooth process.

Because DI(G) Pers 16–16 treated transgender people as medical problems, it did not lead to police investigations and witch-hunts like the LGB ban. Still, there were cases when medical or other authorities suspected members of being transgender and investigated. 'J' served in the Navy from 2003 to 2006 and then transferred to the RAAF, where he is still serving. For years J lived as a lesbian woman but was internally questioning his gender identity. For medical reasons, J needed to have a hysterectomy, but Defence doctors were suspicious that really this was a sly way to have gender affirmation surgery (historically doctors often sought to perform hysterectomies on trans men even though they are medically unnecessary and most trans men do not desire them). The doctors would not let J have the surgery until he saw a psychiatrist. The psychiatrist kept subtly raising the issue of gender transition, and J recalls:

> It hadn't even occurred to me, at the time, to have a sex change. It didn't interest me. Because of the way I presented myself—a lesbian with short cropped hair and boyish clothes—a rumour was started by the medical staff—that I was somehow seeking a hysterectomy as a prelude to a gender transition. Being told you have a mental problem and dismissing my physical problem, and being

> coerced into take anti-depressants as an ultimatum for surgery, took its toll. I was concerned for my career, and it affected me in many ways for years to come.

J had to demand a second opinion and eventually was authorised to have the surgery, although the previous opinion remained on file and caused future dramas. Years later, J was encouraged to put in a complaint over the harassment he suffered from numerous medical staff and received a formal apology, but from the one GP who had actually been supportive.[71]

The years of DI(G) Pers 16–16 left transgender service members with the double burden of struggling with their gender identity while not being able to seek support through the normal channels of doctors, psychologists, psychiatrists or counsellors. When they felt comfortable in private, some transgender service members would dress. Interview participants all describe not only bouts of depression but also being scared to speak about it or to seek help. Amy Hamblin was already seeing a psychologist for PTSD she had acquired during a tour in Iraq. She had been struggling with her gender identity for years and states: 'For the first time in my life I found that psychiatrists and psychologists can actually be helpful. Why don't we deal with this transgender issue? OK. Well, we know that they're good for keeping a secret.'[72]

Summarising her situation, RAAF Squadron Leader Cate Humphries recalls: 'It [gender incongruence] was interfering with me, it was making me depressed, I needed to talk to people, so in 2007 I went to a GP … and the GP referred me to a psychiatrist in Adelaide where I was able to at least start talking and exploring.' Humphries took a big risk and came out to her commanding officer and senior medical officer. She explains: 'Defence was very specific. If you were receiving medical treatment you were to tell Defence … So my integrity meant that I did have to tell the medical officer. And as I said, the medical officer was, thankfully, amazing and protected me from the adverse effects.'[73] Other service members who came out were not so lucky, and it would be two of these members who challenged the transgender ban.

Challenging DI(G) Pers 16–16

Just before Christmas 2009, Captain Bridget Clinch drafted a letter to her colleagues. A few months earlier, a specialist had diagnosed Bridget with gender incongruence and she had come out as transgender to her supervisor and his supervisor. Clinch's letter explained what being transgender meant, the transition process and how overseas militaries such as Canada had successfully managed transgender service. She remembers the response from her colleagues:

'When I came out at work, what was really interesting was like, there was this wave of support … and that was, I guess, really heartening.'

Clinch began taking hormones in November 2009, but in December 2009 she received notification that she could not dress as a woman at work while legally she was still male.[74] In February 2010, Clinch received notice that her medical classification had been downgraded and, 'As a result of being medically unfit for deployment and considering your treatment may be for a prolonged period, regrettably I must inform you that you are medically unfit for further service. Consequently, it is my intention to seek your discharge on medical grounds.'[75] The medical discharge meted out to Clinch exposes why DI(G) Pers 16–16 was essentially a transgender ban: being transgender automatically meant medical downgrade, and that downgrade meant unfit for service.

Clinch appealed to the Medical Classification Review Board and lodged a complaint with the Australian Human Rights Commission. She prepared a brief that challenged the assumptions underpinning each section of DI(G) Pers 16–16, concluding: 'It seems to me that Defence is of the belief that it is above anti-discrimination law. No employer regardless of who they are in this country has the right or the authority to terminate someone's employment based on the fact that they are transsexual.'[76] At the time, gender identity was not a protected category in Commonwealth law (it would be added to the Sex Discrimination Act in 2013), but she is correct that there were protections under state and territory laws. Clinch's Human Rights Commission complaint alleged violations under the Disability Discrimination Act (calling her diagnosis of gender identity disorder a disability), the Convention on the Rights of Persons with Disabilities and the International Covenant on Civil and Political Rights (the same international law under which the LGB ban had been challenged nearly twenty years earlier). In July 2010 the ADF formally withdrew Clinch's termination, but by then the entire process had taken a significant toll on her mental health and well-being.

At the same time that Clinch was challenging DI(G) Pers 16–16, RAAF member Amy Hamblin was also confronting the policy. She had been serving since 2001 in logistics and had already been diagnosed with PTSD following a 2006 tour of duty in Iraq. Hamblin did not voluntarily come out as transgender, but rather was caught by a mate in her private residence wearing a nightgown. The mate reported Hamblin, and she had to front up to the wing commander. She remembers thinking:

> I'm not going to refute it. I'm not going to lie; I'm going to be honest. And so from that moment on I said, 'Yes, Sir, after hours I live as a woman', and he goes, 'Oh, OK, are you aware of this

> policy [DI(G) Pers 16–16]?' … and I said, 'Well, OK, I'll wait for someone to officially challenge me on it and then I will fight it.'[77]

Hamblin challenged DI(G) Pers 16–16 through the RAAF chain of command and prepared a legal case in case it were necessary.

The two challenges, especially the external one through the Human Rights Commission, put pressure on the ADF to reconsider the entire transgender ban. In May 2010, Clinch received a notice from a brigadier acknowledging her appeal and which also stated: 'Your request for the DI(G) Pers 16–16 Trans-gender Personnel in the Australian Defence Force to be subject to immediate cancellation and review has been initiated, and the policy is in the process of being cancelled.'[78] Clinch's termination notice was withdrawn in July 2010, and at last, on 2 September 2010, the ADF issued a DEFGRAM memo announcing the repeal of DI(G) Pers 16–16 and advising commanders to 'manage Australian Defence Force (ADF) trans-gender personnel with fairness, respect and dignity in accordance with DI(G) PERS 50–1—*Equity and Diversity in the Australian Defence Force* and existing Defence medical review provisions'.[79] The DEFGRAM promised that an interim policy on transgender service would be in place by December 2010—a deadline that it did not meet—leaving a policy vacuum and a long period of uncertainty.

Lifting DI(G) Pers 16–16 was only the first step, and both Clinch and Hamblin ran into obstacles as they tried to continue their transitions. For instance, Hamblin required medical reports for everything ranging from obtaining a new women's uniform to acquiring hormones from the base pharmacy. Clinch reported difficulties obtaining a women's uniform, regulations around hair length and access to women's toilets. In a report responding to Clinch's complaints, the Inspector General of the Australian Defence Force acknowledged that because there was no policy on transgender personnel, 'Almost every issue needed to be discussed and researched before decisions were made.'[80]

The biggest uncertainty was whether the ADF would pay for transgender members' gender affirmation surgeries. This question and the entire issue of transgender service went public when Clinch and her wife appeared on Channel Seven's show *Sunday Night* and did an interview for the magazine *New Idea*. Gender affirmation surgery became the main hook for other media reports, with headlines like 'Defence to foot bill for sex change surgery' and 'Military funds for sex change'.[81] Two days before the *Sunday Night* story, the Army advised that it would pay for Clinch's and other members' gender affirmation surgery. Major General Craig Orme said in his *Sunday Night* interview: 'Surgical procedures and the medical treatment that any member of the

Australian Defence Force gets is paid by the taxpayers … We spend a great deal of money training them, and the investment we make in our people, we believe, is well worth the recovery of providing medical support to them.'[82] Although it took more administrative hurdles as outlined above, ultimately the ADF did pay for Bridget Clinch, Amy Hamblin's and other transgender members' medical bills, including for hormones and surgery.

Ongoing difficulties: policy vacuum

The policy vacuum post-DI(G) Pers 16–16 meant that commanding officers had little guidance on how to deal with transgender personnel, leading to variable experiences. Donna Harding's coming out was relatively smooth, possibly because she is a nurse with postings that included the Army School of Health and Army Headquarters. In August 2012 she notified her commanding officer that she was transgender, and she remembers the colonel's response: '"You're the second one this week." He was talking about Cate McGregor.' Harding's CO worked constructively to communicate her situation to her co-workers. She drafted a letter for her unit that explained her personal story and why she was transitioning. Harding was absent when her CO read the letter because she wanted the group to feel able to express any thoughts or questions without embarrassment. When Harding returned to work after transitioning, her colleagues were affirming: 'I came back to work as Donna. It was as if I'd been on holidays; no mis-gendering, no mis-, nothing like that.'[83]

As mentioned above, RAAF officer Catherine Humphries had already come out to her CO. Even after the repeal of DI(G) Pers 16–16, Humphries still did not transition because she was in a combat role not open to women. In September 2011 the Australian Government ordered the ADF to open all combat roles to women over the next five years. Humphries' role as a ground defence officer, guarding air installations in Australia and abroad, was opened to women from January 2013. In late 2012, after confronting depression in her personal life, Humphries finally made the decision to transition. In March 2013 Humphries made a public announcement, and she recalls: 'It took two hours from that announcement in RAAF Base Amberley [Brisbane] to make it to No. 2 Airfield Defence Squadron in the field in Shoalwater Bay [central Queensland] … gossip travels fast.' Humphries believes that by being open she 'took power away from any under-the-table rumours'.[84] Importantly, both Humphries and Harding only took short periods of leave when they were undergoing medical transition; otherwise, they continued to work.

Both Humphries and Harding were high-ranking officers: a RAAF squadron leader and an Army major respectively. Humphries observes: 'I have the benefit of being, I suppose, trained and in an alpha-style career field, and

I'm of a reasonable rank. I can make things happen that a low rank … someone who's more uncertain or doesn't know the system—they can't make it happen.'[85] Other ranks did not have the same influence and consequential sense of control over their workplace treatment as others. Harding describes one private who 'had an absolute shit of a time', being posted to a new unit after she transitioned. Harding asserts that this undermined perceptions of the private's abilities and set her up for close scrutiny.[86]

Transgender people going through Defence Force Recruiting (DFR) also confronted challenges because of the lack of clear policy. Dana Pham was in the process of transitioning when she applied in early 2011 to join the ADF. The ADF rejected her first application because, according to the letter, she had been on antidepressants within the previous year. Pham sought further information and remembers the response: 'There was this one thing he said that really irritated me, which was—oh, it was something along the lines of unless I've had my gender reassignment surgery I can't be considered.' Pham appealed the rejection and sent letters of support from her psychiatrist explaining the reason for the antidepressants. She quotes directly from the DFR medical's response to her appeal: 'You being a woman of transsexual background does not in itself disadvantage you … The Chief Medical Officer, DFR, considers that the medical implications of your transitioning gender status are significant. Further information is required in order to assess your fitness for military service.' Pham persisted with DFR and from April 2012 was continually supplying reports from her psychiatrist and endocrinologist. As she kept dealing with new doctors, she felt that every time she 'was giving this doctor a 101 on the medical management of transgender people'.[87]

For every issue that Pham addressed, doctors came back with new ones. She describes another letter dated April 2012, which raised concerns about hormone medication: 'This doctor's concern was if we cut off your medication for whatever service reason we need to know: 1. how quickly your secondary sex characteristics will redevelop as a male, and 2. the psychological impact of that … Essentially what they're looking for is a stable hormone regime ensuring that there are no post-operative complications.' At that stage Pham had not yet decided whether she wanted gender affirmation surgery, and this links to a key point lost in many of the ADF policy frameworks around transgender health and service: not all transgender people want or need gender affirmation surgery. Moreover, gender affirmation surgery is more appropriately classified as surger*ies*, as some transgender people might desire some surgeries but not others. Pham ultimately did have gender affirmation surgery in late 2012, and she confesses that the main reason was to get through DFR—highlighting the ways policies could affect health decisions best reserved for the transgender person and their doctors. Pham succinctly says: 'I don't regret my gender transition. I do regret making a medical decision just to improve, just to play the game with DFR.' Finally, DFR Medical approved Pham for admission in July 2013. At last, in May 2014 Pham received her offer letter and finally went to officer training school and subsequently trained as a Personnel Capability Officer.[88] Dana Pham's enlistment process took more than three years, most delays being due to the lack of policy or guidelines around supporting transgender service.

In November 2011 the Department of Defence finally released a document entitled 'Understanding Transitioning Gender in the Workplace'. It explained some of the key medical and social understandings of transgender and raised some of the important considerations for commanders of Defence members who were transitioning: change of name, uniforms, use of pronouns and toilet access. The document also said that 'each member will be managed on a case-by-case basis'.[89] As Amy Hamblin notes, however, this document was inadequate because it did not effectively explain how to support transgender members in the ADF: 'And so it left everyone in limbo and if you've ever understood anything in Defence, everything that's in Defence is black and smudge. If it isn't in black and smudge you don't do it.'[90]

Amy Hamblin began working with a RAAF reservist from the diversity directorate to develop a transcentric guide that could support both transitioning members and commanders. In April 2013, the Air Force Workforce Diversity Directorate published the *Air Force Diversity Handbook: Transitioning Gender in Air Force*. This is one of several diversity handbooks that the RAAF published (including two others on sexuality). The transitioning gender handbook is particularly detailed because transgender pioneers Amy Hamblin, Donna Harding and Catherine Humphries all contributed on the basis of their own transition experiences. As Hamblin explains: 'We realised that the actual transition is very unique to the individual ... We need to make sure that what we write is very generic and that the people who are empowered are the people who need to be empowered for the associated information.'[91] The handbook consistently emphasises the distinctiveness of every member's transition journey and often uses the second person with statements like 'your unique situation'. It outlines terminology and how to navigate some of the most obvious gendered aspects of RAAF and civilian life, including passports, identification documents and uniforms. The handbook also includes a few strategies for coming out and the strengths and weaknesses of different options, such as whether to transfer units. There are tips for transitioning, such as finding a mentor in the affirmed gender, how to take leave during transition and how to manage difficult questions. Advice to commanding officers explains how to protect members' privacy and how to confront bullying, harassment and transphobia. Finally, the guide has four annexes to support transitioning members: a gender transition support plan, a roadmap for discussions with commanding officers, a sample letter explaining their transition to commanding officers and a sample letter to co-workers.[92]

Transgender Defence members in the three services have supported each other in other ways, with officers especially playing leadership roles. Many transgender service personnel are members of DEFGLIS, which now has at least one board position reserved for a transgender or intersex member. When Catherine Humphries was on the DEFGLIS board, she set up a new closed Facebook group for transgender Defence members for 'people to ask questions and particularly for transgender, and people thinking, or considering their gender identities'.[93] By 2016, the Facebook page had approximately twenty-five members, and as of June 2020 there were forty-seven members.

Superior officers have, where possible, offered support to more junior ranks and to transgender people who have just joined the ADF. For instance, Dana Pham was a bit 'panicked' before starting RAAF Officer Training School, so she reached out to a more senior transgender officer. That officer explained what Officer Training School would be like and, with Pham's permission, notified some of the staff in advance about her being transgender. The staff assigned the room allocations so that Pham did not have to share an en suite.[94] Joel Wilson was a RAAF officer cadet at ADFA when he began transitioning in 2013. He was the first ADFA cadet to transition and ran into many difficulties from the start, as outlined in chapter 7. He did, however, find support from both Donna Harding and Catherine Humphries via DEFGLIS.[95] This peer support and information-sharing would prove vital as more transgender people joined the ADF or Defence members announced their intention to transition.

Conclusion

The long histories of transgender and intersex military service in many ways mirror their broader histories in Australia. Public discourses and social stigmas around the body, gender and sexuality often rendered transgender and intersex people invisible and silent. For transgender people who transitioned before the 1990s, the general expectation, perpetuated by the medical profession, was that they should have gender affirmation surgery and then disappear quietly into society. For many older transgender service members who left the forces and transitioned, this was the case. By the 1990s, though, as a new wave of transgender activists pushed for anti-discrimination and acceptance, the ADF began designing policies that explicitly barred people from transitioning gender while serving. In many ways, the implementation of DI(G) Pers 16–16 in April 2000 mirrors the introduction of the tri-service LGB ban in 1974: it was a reaction to growing visibility, anti-discrimination laws and activism in civil society. It would take brave pioneers to challenge the transgender ban, and again it was external pressure from the Australian Human Rights Commission that forced the ADF to review its policy and repeal the ban.

Since 2010, transgender service members have slowly come out and transitioned in the ADF. The courageous individuals who were among the first wave of openly transgender service members had to navigate a policy-less space where they were regularly setting precedents, had to be persistent and themselves designed guides and provided peer support for newer and lower-ranked transgender members. As chapter 7 will show, the experiences of transgender members have improved over time, but progress has been uneven. New understandings about gender diversity and the enlistment of the first non-binary members in 2016 and 2017 would further test the ADF's commitment to transgender inclusion amid an increasingly hostile political climate. Moreover, the medicalisation of transgender policy became even more problematic when the ADF eventually adopted a new policy for managing transgender service members in 2015. Indeed, just as trans and gender-diverse people were finding their place in the ADF, the institution would unwittingly find itself tied up in the new culture wars.

Notes

1 See 'Watch David Morrison's 2013 video in which he tells troops to respect women or "get out"', ABC, 14 June 2014, https://www.abc.net.au/news/2014-06-14/david-morrison-youtube-video/5524058, retrieved 2 January 2020.

2 'Lieutenant Colonel Cate McGregor', *One Plus One*, ABC, 5 July 2013, https://www.abc.net.au/news/programs/one-plus-one/2013-07-05/one-plus-one-lieutenant-colonel-cate-mcgregor/4802564, retrieved 2 January 2020.

3 See Stephen Kerry, 'Australian News Media's representation of Cate McGregor, the highest ranking Australian transgender military officer', *Journal of Gender Studies* 27, no. 6 (2018): 683–93.

4 'Lieutenant Colonel Cate McGregor', *One Plus One*.

5 George R. Brown, 'Transsexuals in the military: Flight into hypermasculinity', *Archives of Sexual Behavior* 17, no. 6 (1988): 527–37.

6 George R. Brown and Kenneth T. Jones, 'Mental health and medical health disparities in 5135 transgender veterans receiving healthcare in the Veterans Health Administration: A case-control study', *LGBT Health* 3, no. 2 (2016): 127; Mildred L. Brown and Chloe Ann Rounsley, *True Selves: Understanding Transsexualism—for Families, Friends, Coworkers, and Helping Professionals* (San Francisco: Jossey-Bass, 1996), pp. 79–80.

7 Joshua Polchar et al., *LGBT Military Personnel: A Strategic Vision for Inclusion* (The Hague: The Hague Centre for Strategic Studies, 2014); Riseman, 'Transgender policy in the Australian Defence Force', pp. 141–54; Adam F. Yerke and Valory Mitchell, 'Transgender people in the military: Don't ask? Don't tell? Don't enlist!', *Journal of Homosexuality* 60, no. 2–3 (2013): 436–57.

8 Kelly Jacob Rawson and Cristan Williams, 'Transgender*: The rhetorical landscape of a term', *Present Tense* 3, no. 2 (2014): 1–9.

9 Arguably the first major text about trans men was Leslie Martin Lothstein, *Female-to-Male Transsexualism: Historical, Clinical, and Theoretical Issues* (Boston: Routledge & Kegan Paul, 1983).

10 Susan Stryker, *Transgender History: The Roots of Today's Revolution*, 2nd edn (New York: Seal Press, 2017), p. 1.

11 'GLAAD media reference guide—Transgender', https://www.glaad.org/reference/transgender, retrieved 15 May 2019.

12 'What is intersex?', Intersex Human Rights Australia, https://ihra.org.au/18106/what-is-intersex/, retrieved 2 January 2020.

13 Magnus Hirschfeld, *Transvestites: The Erotic Drive to Cross Dress*, trans. Michael A. Lombardi-Nash (New York: Prometheus Books, [1910] 1991).

14 Harry Benjamin, 'Transsexualism and transvestism as psycho-somatic and somato-psychic syndromes', in *The Transgender Studies Reader*, ed. Susan Stryker and Stephen Whittle (New York: Routledge, [1954] 2006), pp. 45–52.

15 Victoria Haskins, 'The girl who wanted to go to war: Female patriotism and gender construction in Australia's Great War', *History Australia* 14, no. 2 (2017): 169–86.

16 Jenny, 'Did you hear the one …', *Transceiver* (WA), December 1996, p. 7.

17 Smaal, *Sex, Soldiers and the South Pacific*, p. 84.

18 'Ex-RAAF sergeant will become a woman', *Truth* (Adelaide), 3 September 1955; 'Ex-RAAF sergeant will become a woman', *Truth* (Sydney), 28 August 1955, pp. 1–2.

19 Neil Buhrich, 'A heterosexual transvestite club: Psychiatric aspects', *Australian and New Zealand Journal of Psychiatry* 10, no. 4 (1976), p. 333.

20 'Why they do it', *Independent* (Perth), 14 November 1976.

21 Susan Williams, 'Life's journey for Sue', *Feminique* (Sydney), issue 11, 1976, p. 13.

22 'Profile: Lynette Jones', *Seahorse Newsletter* (Victoria), February 1983, p. 8.

23 Terri, 'Naval award for first vessel into the Monte Bellos', *Transceiver*, October 2005, p. 5.

24 Marie-Desirée D'Orsay-Lawrence, Senate, Legal and Constitutional References Committee, 'Reference: Sexuality discrimination inquiry', Adelaide, 16 August 1996, p. 514.

25 Kaye Rona Ellis, 'I am a woman', *Seahorse Scene* (Victoria), no. 5, 1977, p. 12.

26 Muriel Reddy, 'When a man becomes a woman', *Sunday Age*, 9 September 1990, p. 1.

27 Kerryn Phelps, 'I am woman', *Daily Telegraph* (Sydney), 21 June 1993, p. 30, in Queensland AIDS Council transgender archive, notebook 29, recently donated to ALGA; Kerryn Phelps, 'The airman and his wife', *Herald Sun* (Melbourne), 21 June 1993, p. 23.

28 Gina Mather, Senate, Legal and Constitutional References Committee, 'Reference: Sexuality discrimination inquiry', Brisbane, 1 October 1996, p. 790.

29 Cummings, *Katherine's Diary*, p. 55.

30 This claim may be tied into numerous mythologies around National Service and Vietnam. See Ben Morris and Noah Riseman, 'Volunteers with a legal impediment: Australian National Service and the question of overseas service in Vietnam', *History Australia* 16, no. 2 (2019): 266–86.

31 'Penelope', interview with Noah Riseman for transgender history project, 13 October 2018, Melbourne.

32 Quoted in Riseman, Robinson and Willett, *Serving in Silence?*, p. 91.

33 'She'll be a shock for the sar' major', *Truth* (Melbourne), date unknown, in Queensland AIDS Council transgender archive, notebook 1.

34 'Is he in the Army?', 'Army check for Les Girl' and 'Mini for Army check', dates and newspapers unknown, in Queensland AIDS Council transgender archive, notebook 1.

35 Corporal K.A. Knight, 'Theft of WRAAF undergarments from RAAF Base Pearce WRAAF Barracks', 13 December 1971, NAA PP479/1, 705/1/P1 PART 4.
36 NAA ST5817/1, 705/24 P1 PART 1; NAA ST5533/1, 705/15/P1 PART 1.
37 Flying Officer E.S. Hulbert, 31 January 1973, NAA ST5817/1, 705/24 P1 PART 1.
38 Karen Patterson, 'Fighting for recognition', *Sun* (Brisbane), 14 December 1990, available from RMIT Social Policy Archive: Social Health Press Clippings Collection, Melbourne eScholarship Research Centre, Series 1—Social Health Press Clippings Collection, 'Transvestites/Transsexuals 7/90—'.
39 Phelps, 'I am woman', p. 30; Phelps, 'The airman and his wife', p. 23.
40 'Kate', interview with Noah Riseman, 31 July 2015, Melbourne.
41 'Kate', interview.
42 Michael Cox, 'I want to be a "she": Ex-digger fights for sex change', *Truth* (Melbourne), 1 January 1983, p. 11.
43 Greg Mansfield, 'Why Allan disappeared, and returned as Vicki', *Advocate* (Burnie, Tas), 9 September 1977, available from RMIT Social Policy Archive: Social Health Press Clippings Collection, 'Transvestites/Transsexuals 5/76—12/79'.
44 N. Pollock service record, 0115708.
45 Catherine Heal, interview with Noah Riseman, 28 July 2017, Melbourne.
46 N. Pollock service record, 0115708.
47 Catherine Heal, interview.
48 'Helen', interview with Noah Riseman for transgender history project, 25 July 2019, Melbourne.
49 'Intersex population figures', Intersex Human Rights Australia, https://ihra.org.au/16601/intersex-numbers/, retrieved 7 January 2020.
50 'Darlington statement', 10 March 2017, https://darlington.org.au/statement/, retrieved 7 January 2020.
51 Ana Lúcia Santos, 'Beyond binarism? Intersex as an epistemological and political challenge', *RCCS Annual Review*, no. 6 (2014): 127.
52 Stirling, *So Different*.
53 'Gina', interview with Noah Riseman, 30 May 2017, Sydney.
54 'Gina', interview.
55 'Gina', interview.
56 'Gina', interview.
57 Ben Robertson, 'I am half-man, half-woman', *Woman's Day*, 3 August 1992, pp. 12–13.
58 Ibid.
59 Ben Robertson, 'Tony in army sex dilemma', *Herald-Sun*, 3 July 1992, p. 3; Robertson, 'I am half-man, half-woman', p. 13.
60 NAA B2458, 331048; *Australian Military Regulations*, 19 July 1990.
61 'Gina', interview.
62 Noah Riseman, 'Transgender inclusion and Australia's failed Sexuality Discrimination Bill', *Australian Journal of Politics and History* 65, no. 2 (2019): 259–77.
63 Senate Legal and Constitutional References Committee, Sexuality Discrimination Inquiry, submission no. 172, Department of Defence, 5 August 1996.
64 Ibid.; Commodore Jim O'Hara, Senate Legal and Constitutional References Committee, 6 August 1996, p. 8.
65 Piers Akerman, 'How to tell a man from a woman', *Sunday Telegraph* (Sydney), 11 August 1996, p. 125.
66 Bruce Ruxton, 'Three cheers for Piers', *Sunday Telegraph*, 8 September 1996, p. 54.
67 DI(G) Pers 16–16, 'Trans-gender Personnel in the Australian Defence Force', 20 April 2000.

68 Ibid.
69 Ibid.
70 'Mitch', interview with Noah Riseman, 9 August 2018, Adelaide.
71 'J', interview with Noah Riseman, 14 November 2016, Canberra.
72 Amy Hamblin, interview with Noah Riseman, 11 September 2015, Melbourne.
73 Catherine Humphries, interview with Noah Riseman, 11 August 2015, Canberra.
74 Bridget Clinch, interview with Shirleene Robinson, 30 July 2015, Brisbane.
75 Colonel P.M. Nothard, Director Officer Career Management—Army, to Bridget Clinch, 26 February 2010, available from Bridget Clinch Human Rights Commission case documents.
76 Bridget Clinch, 'Brief on the reasons why DI(G) PERS 16–16 Trans-gender Personnel in the Australian Defence Force is not a valid reference', March 2010, available from Bridget Clinch Human Rights Commission case documents.
77 Amy Hamblin, interview with Noah Riseman, 26 August 2015, Melbourne.
78 Brigadier M.W. Hall, Commander Career Management Army, to Bridget Clinch, 5 May 2010, available from Bridget Clinch Human Rights Commission case documents.
79 DEFGRAM NO 553/2010, 'Trans-gender personnel in the Australian Defence Force', 2 September 2010.
80 Letter from Inspector-General of the Australian Defence Force to Bridget Clinch, n.d., courtesy Bridget Clinch.
81 Ian McPhedran, 'Defence to foot bill for sex change surgery: The Digger trapped inside a man's body', *Advertiser* (Adelaide), 8 November 2010, p. 11; Mark Dunn, 'Military funds for sex change', *Courier-Mail*, 29 October 2010, p. 28.
82 'An Australian soldier's bravest moment', *Sunday Night*, 7 November 2010, Seven Network.
83 Donna Harding, interview with Noah Riseman, 10 August 2015, Canberra.
84 Catherine Humphries, interview.
85 Catherine Humphries, interview.
86 Donna Harding, interview.
87 Dana Pham, interview with Noah Riseman, 24 May 2015, Sydney.
88 Dana Pham, interview.
89 Department of Defence, 'Understanding transitioning gender in the workplace', 10 November 2011.
90 Amy Hamblin, interview, 26 August 2015.
91 Amy Hamblin, interview, 26 August 2015.
92 Air Force Workforce Diversity, *Air Force Diversity Handbook: Transitioning Gender in Air Force*, April 2013. An updated version of the handbook was published in July 2015.
93 Catherine Humphries, interview.
94 Dana Pham, interview.
95 Joel Wilson, interview with Noah Riseman, 15 May 2018, Canberra.

Chapter 7

A new ADF?

In 2012 the Chief of Army, Lieutenant General David Morrison, told the conservative-leaning Sydney Institute that twenty-five to thirty years earlier, 'the reaction to people of a different sexual orientation [in the ADF] would have been as almost insurmountable … yet now of course it isn't an issue and nor should it be. And we have many very proud gay and lesbian soldiers, airmen, airwomen, sailors serving in our ADF.'[1] In 2010 the ADF reformed its policies to allow open transgender service. Since then, the ADF has publicly affirmed itself as a welcoming and inclusive environment for LGBTI personnel and has actively pursued the inclusion of diverse sexualities and gender identities, recognising the way that these members strengthen ADF capabilities. The 'Defence Diversity and Inclusion Strategy 2012–2017' even includes LGBTI persons as a group requiring priority attention.[2]

This chapter explores the position of LGBTI service personnel in the contemporary ADF. Although the organisation has taken significant strides to support LGBTI members, there are still areas where reform is needed. These include improvements that address the needs of transgender, gender-diverse and non-binary members. The ADF also continues to grapple with supporting members living with HIV. Yet, support from senior ADF leadership in recent years has been important to advance the position of LGBTI service personnel. The advocacy organisation DEFGLIS has also played a critical leadership role in bringing LGBTI members together and creating a more inclusive climate for LGBTI service personnel. Evolving attitudes within the ADF have been apparent through symbolic gestures such as permitting ADF members to march in the Sydney Gay and Lesbian Mardi Gras in uniform since 2013 and supporting the annual Military Pride Ball since 2015. These strides have been important as the ADF and LGBTI service in it have unwittingly become pawns in culture wars waged by conservative politicians and pundits.

Sexuality-based prejudice in the ADF

The journey to LGBTI inclusion in the ADF has been a drawn-out and uneven process. In 2013, 32 per cent of LGB Defence members who responded to an informal survey reported that they hid their sexuality at work. Nine per cent of respondents reported harassment and discrimination within the ADF based on their sexuality. On the positive side, however, 59 per cent of survey respondents reported being open about their sexuality and it having no impact on their work environment.[3] Although the survey had a small sample size, it reflected broader patterns also evident in our oral history interviews: there has been significant progress in the ADF, but that does not mean that it is an environment free of prejudice on the grounds of sexuality.

Fifty-one of our oral history participants were serving in the regular or reserve forces at the time of interview: thirty gay or bisexual men; ten lesbian, queer, bisexual or gay women; and eleven transgender and non-binary people. While almost all of these interview participants are open about their sexuality or gender identity, many of them chose pseudonyms because they do not want to be readily identifiable. It is not possible to quote from all of them, but sample extracts show affirming experiences for LGB Defence members:

> 'What people say, it's like, "We don't care that you're gay. It's you as a person." … I was culturing both worlds [gay and Army]. I had gay guys going, "I have never had so many straight friends and they're so hot to look at." You know … they loved the attention, the Army guys.' (Scott Amos, Army)[4]

> 'I've been pretty lucky, but also, when you're a little bit more senior as well, you're more in a position where people aren't going to come and be horrible to you directly and you sort of feel that you're more in a shaping position. Like, when I was doing my deputy posting on *Sydney*, one of the junior sailors came up to me and said, "Ma'am, you're out, aren't you?" He must have, probably had duty [*sic*] at some time. I was like, "Yeah, yeah, I'm totally out if you need a hand or anything. Yeah, that's fine." He found it a good thing to have one of the officers being out.' ('Linda', Navy)[5]

> 'We have a really supportive community here [ADFA]. We have a group on Facebook where everyone chats, and we have social

> events throughout the year. We've just finished a dinner … the International Day Against Homophobia/Transphobia [IDAHOBIT] … this massive dinner … It's the dinner with the most external Defence guests at ADFA.' (Connor Haas, then RAAF ADFA cadet)[6]

Discussing a conversation during basic training at Kapooka when the supervising corporal asked the recruits what they missed:

> 'Oh, chocolate, beer, all the usual things, and one of the girls said, "Hugs. I just really miss hugs", because you're not allowed to touch each other. And [the corporal's] like, "Well, you can hug Billett. He's safe." And I've just gone bright red, because all my section knew [I'm gay] but these other guys from the other sections didn't, and they look at me, and they're: "What are they talking about?" I said, "I'm gay", and they went, "Oh", and nobody cared. But then any of the girls that wanted hugs had to come to me. But then it came around one of the guys from one of the other sections was, "I miss hugs, too", and Corporal looked at me and went, "Do you mind?" I went, "Nah", so I got to hug him as well.' (Graeme Billett, then Army [now Navy])[7]

Although affirmation was evident, all interview participants, even those who reported overwhelmingly positive experiences, had witnessed homophobic comments. Several interview participants note the common use of the expression 'that's so gay' or the words 'poofter' and 'faggot'. RAAF member Neal Fischer explains from his experience: 'If the boss gives you an order that's, you know, not to everyone's satisfaction, you will still hear people say to this day, "Oh, that's gay", or words to that effect. Sometimes it escalates to "What a faggot thing to do".'[8] Importantly, however, many LGB members feel empowered to call out this casual homophobia. Rob Brown, who served in the RAAF from 2002 to 2020, observes: 'People who don't know, who make snide jokes at work and in the lunch room: "Bloody poofs, are this and …" And sometimes I'll pull them up on it. Depends on who it is and what the attitude is. If they're being really offensive, I will definitely say something. And I have no problems doing that.'[9] RAAF reservist Becky Green observed a major increase in homophobic comments both in person and during the 2017 marriage equality survey: 'It's now that we're talking about it that people feel enabled to bloody say all this stuff.'[10]

Homophobia could also be targeted at individuals, although often in subtle ways. RAAF officer 'Diego' came out to his colleagues during pilot training and found it to be a liberating experience that took a weight off his shoulders. He continued to be close to his colleagues during training for a particular aircraft a few years later, but several older instructors were being especially hard on him. He remembers one instructor told him: 'Mate, it's like you just don't fit in.' Diego did not understand why the instructor would say this, given he had an active social life with his co-pilots. On reflection, Diego realised: 'They apply this heteronormative veil on top of their expectations, and they didn't see what they wanted to see, even though I was socially connected with my group. They were like, "But he's not like us, so he doesn't fit what I'm expecting to experience from a co-pilot."'[11] These trials hurt Diego's self-esteem and disrupted his career trajectory; that said, new postings serving under affirming officers rebuilt his confidence, and he is now thriving.

Lesbian and bisexual women report experiencing homophobia and sexism in different ways from men. As previous chapters have outlined, sexism has been a constant factor women have encountered as long as they have served. In 2013, 'Larissa', who identifies as bisexual and queer with a female partner, joined the Army as a combat engineer. Socially, she found some of the older generation of men, usually older than fifty, expressed outdated views about gender roles and sexuality. She states:

> Certain places within Defence—it's not the kind of place any self-respecting woman would want to spend too long with team-mates like that that don't really respect what you can do on your own. [They] think that the only reason you've got to where you are is because there's the double standards that have made it easier for you to get there.[12]

'Cooper', a gay airwoman in the RAAF, experienced harassment while working in a non-traditional role for women. Some co-workers used to bully her by making comments about gays and women in the ADF. Cooper is explicit about the intersections of homophobia and sexism: 'But then also I don't think it was just a homophobia thing in that [treatment]. I think they had their own issues over the fact that I was a chick.' Cooper also experienced other forms of casual sexism: regularly she would be assigned to do paperwork instead of her male colleagues, and she would be the one doing all the cleaning at the end of the week while her male co-workers watched.[13]

Other women have been conscious of the way they are treated differently from gay men and straight women. Able Seaman 'Lexa' works well with male sailors because they often see her as one of the 'guys' who enjoys joking around. But she also observed on one ship deployment: 'A few of the guys initially tried very hard to get to know me, and I'm not naive. They wanted to know what I was doing, what I liked, my favourite colour, la, la, la. And the second they found out I was gay and I wouldn't sleep with them, they stopped talking to me entirely.'[14] Both Larissa and Lexa believe that gay men are more likely to encounter prejudice on the grounds of their sexuality within their respective services. Larissa says of the Army: 'I've heard examples of some gay men that have had a bit more of a hard time, usually from the older guys with rank',[15] while Lexa comments about the Navy: 'They ostracise gay guys. They don't really want a bar of them. They're like, "He's weird. He's gay." But it's all right for a straight guy to slap another straight guy on the ass.'[16]

Leading Seaman Adam Pritchard's experience supports those observations and also reveals that the ADF often did not manage complaints well. Shortly before his discharge in 2006, he was posted in the investigations section of the Naval Police at HMAS Cerberus. His chief petty officer regularly made homophobic comments like 'Faggots are disgusting'. This escalated, and the warrant officer, too, would make homophobic and sexually inappropriate remarks that targeted Pritchard. He remembers the warrant officer yelling down the corridor: 'Leader, are you waiting at my office to give me a blow job?' On another occasion, the warrant officer said something to the effect of 'Yeah, I know what you used to do when you were on board. Get on your hands and knees and blow all the boys.' With the support of a female colleague, Pritchard finally lodged a grievance over the bullying and harassment. The investigator was, according to Pritchard, a homophobic Naval Reserve lawyer, and the final report was a slap on the wrist with a note about the workplace culture needing to change. Years later, Pritchard filed a claim with the Defence Abuse Response Taskforce (DART) over this bullying and the ADF's poor handling of the complaint. Like others who went through DART, Pritchard was pleased with the outcome: they affirmed that he had been bullied and the complaint mishandled. Defence even offered to reopen the investigation, but Pritchard did not want to revisit that difficult part of his past.[17]

More serious reported cases of homophobia are rare, but they still occur. In August 2010, the ADF Investigative Service was notified about offensive Facebook pages that named and vilified homosexual personnel. The media reported that dozens of soldiers had reportedly interacted with the pages that targeted five ADF personnel for their 'filthy lifestyle decision' and set out to expose 'who is biting the pillow'.[18] One hate page allegedly 'had links to violence and pornographic videos on YouTube depicting homosexuals being executed, superimposed over images of Gallipoli and flag-draped coffins of dead Australian troops'. DEFGLIS provided a statement that the online hate messages were a malicious attack on men and women selflessly serving their country at home and overseas. Stuart O'Brien said, 'I am saddened by this incident and am dismayed that in this day and age, supposed members of the

ADF who pride themselves on possessing the qualities of honour, loyalty and integrity would publicly and viciously attack fellow serving members.'[19] In May 2011, a former member of the ADF was charged over his alleged role in creating the hate page and for email threats sent to an Army major about his sexuality.[20]

Perhaps the most high-profile example of homophobic and transphobic prejudice in the contemporary ADF is the case of Bernard Gaynor. Gaynor had been a member of the Army Reserve from 1997 to 1999, then transferred to the Regular Army and graduated from Duntroon as a commissioned officer in 2002. He continued to serve, including deploying to Iraq, before transferring back to the Army Reserve in 2011. Gaynor has always been open about his conservative Catholicism, and from January 2013 he became a frequent online critic of the ADF's inclusion strategy. Gaynor's blog posts were critical of the ADF's support for LGBTI and Muslim members and especially transgender service. It is not worth repeating the transphobic comments here, but a judge later described them as 'personal and offensive'.[21]

ADF top brass, including the Deputy Chief of Army (now CDF) Angus Campbell, wrote to Gaynor advising him to remove the posts. While Gaynor was entitled to express his views in a personal capacity, the problem was that he was doing so while identifying himself as an Army Reserve officer. When Gaynor refused, the Chief of Army sent Gaynor a 'show cause' letter, and the CDF subsequently issued a termination notice. Gaynor lodged redresses of grievance, which were unsuccessful, and he was formally terminated from the ADF on 11 July 2014. Gaynor subsequently lodged proceedings against the ADF in the Federal Court. He invoked numerous grounds, but the most significant was a claim that the ADF had violated the implied freedom of political speech guaranteed under the Australian Constitution. In December 2015 a Federal Court judge ruled in Gaynor's favour. The ADF appealed, and in March 2017 the Full Bench of the Federal Court overturned the ruling, accepting that the ADF was within its rights to terminate Gaynor. In September 2017 the High Court refused to hear a final appeal.[22] The Gaynor case has been troubling for free speech warriors, yet it sent a strong message across the ADF: the organisation was taking a hard line and would not tolerate Defence members using their positions to express views that contradicted the organisation's values.

Transgender service personnel and transphobia

As more time passed since the repeal of DI(G) Pers 16–16, more serving transgender people transitioned and more transgender people enlisted. Transgender members' oral history interviews describe a mix of positive and negative experiences. One perpetual challenge has been that every transgender member is the 'first', whether that be the first in a particular unit, the first trans man, the first non-binary person, the first ADFA cadet, the first at a base or the first at some rank. Being the first has meant a constant need to educate both peers and chains of command about everything transgender. This can be both an empowering and a frustrating experience for transgender service members.

When we asked interview participants about transphobia, they provided varied responses. (In)visibility was a common theme of our interviews: those Defence members who were not visibly transgender were less likely to report transphobia. Catherine Humphries explains:

> So I was going from a male ground Defence officer, [with a] haircut of five millimetres and starting to appear feminine, so there was a period where it was quite obvious. During that period, yeah—people looking at you, staring at you, making comments behind your back. That certainly occurred. But once that's no longer obvious, it, yeah [stopped]. The ongoing thing I deal with is just the general misogyny with women in Defence.

When questioned further about misogyny, Humphries described how as a female the tone of emails she received were more condescending, questions would more often be directed to lower-ranked males, and senior officers commented on her boobs and legs.[23] Conversely, trans men have reported acquiring male privileges after they transitioned. RAAF member Liam Bone notes, for instance, that he faces fewer assumptions about his physical strength/capability; is more respected when he voices an opinion; is not objectified for his body; can wear anything casual without comment or judgement; and in group conversations, people speak directly to him.[24] Joel Wilson says that he was almost 'inducted' into being one of the guys, and they began to include him in conversations about sexual conquests, pornography and masturbation.[25]

When transgender members encountered transphobia, it was often difficult to distinguish those experiences borne out of ignorance from those borne out of prejudice. Regardless of motivation, the impact was still damaging. Overlaying this were the general bureaucratic obstacles that permeate the entire ADF as an organisation. Many commanding officers, psychologists or doctors were so afraid of making a misstep that they prevaricated or referred matters further up the chain of command. This left transgender members in a state of limbo over matters ranging from health to something as small as whether they could give a presentation at a function. One transgender person was not able to do an oral history interview because their chain of command would not answer their request for approval to participate in our project—even though we had command approval from the Vice Chief of the Defence Force (VCDF) to interview any LGBTI Defence member.

Some service members had more blatant encounters with transphobia. 'Mitch' felt very isolated and was bullied by superiors and peers when he served on Army bases in Melbourne, Canberra and Hobart. The three-year posting in Hobart was especially difficult when Mitch learned that a group of lower-ranked reservists were vilifying him. When he completed his posting in Tasmania, Mitch confronted the other members of his unit and unleashed on them:

> 'I'm a trans man, and you all know this, and I can't believe the hate speech that I've been exposed to.' I just went off, and I said to everybody, 'This uniform, don't you know what this means? We are a family. We abide by a Code of Conduct here', and I said, 'There are some people in this room who just do not deserve to be in uniform. Some of you need to take a seriously good look at yourselves, and I wish I'd come out right from the first minute that I got here because this has been so fucking disrespectful.'

Mitch lodged complaints about the bullying, but the ADF did not handle them well. The investigation report never even used the word 'transgender' but instead sanitised the bullying with lines like: 'There's been instances where they've questioned your decision.'[26]

Another example of poor management mixed with transphobia is from ADFA. In 2013 Joel Wilson was the first transgender person to transition at ADFA. Midway through his first year, Wilson came out to an ADFA chaplain, who was supportive and connected him to a trans-friendly GP. Wilson began seeing psychiatrists, and he finally came out to his divisional officer and gave him the *Air Force Diversity Handbook* and information from DEFGLIS and Canberra's transgender support organisation, A Gender Agenda. Major Donna Harding from DEFGLIS facilitated a training session for the staff at ADFA where anyone could ask questions. Some of the questions were constructive, but Wilson recalls staff posing queries that made him uncomfortable, particularly around whether the ADF would pay for his medical treatment. Wilson came out to his peers at a barbeque at the end of his first year; he recalls: 'Some people came up and were like, "Wow, you're really brave." Most people were like, "Oh, OK, that's weird" and just went on with their day and didn't really interact with me.'[27]

When Wilson returned to ADFA for his second year, the transphobia began in earnest. He recalls one cadet who did not know him saying 'something about how one of the girls couldn't do something because she was a guy, and he's like, "Ha, like why don't you just go get a sex change? Don't you know like it's fucking ridiculous—Defence pays for it. There's someone here getting a sex change."' The rumour mill was very active, saying that Wilson had gone on summer break a female and returned a male. Cadets, some of whom were well meaning, kept asking Wilson how his operations went—even though asking questions about trans operations is highly inappropriate. Wilson grew more socially isolated and his mental health declined. ADFA was inconsistent in whether they treated Wilson as male or female. He aspired to and met the male fitness standards, but there was no flexibility such as giving him extra change time to remove or apply his binder. When he went to the ADFA health centre they misgendered him, giving him female name tags and admitting him into the women's ward. When Wilson began exploring options for top surgery, the ADF doctors kept saying 'no' even though the procedures should have been covered by Defence Health. Feeling isolated and depressed, Wilson made the decision to discharge and leave ADFA.[28]

After he discharged, Wilson learned that he was being investigated by the Department of Defence for defrauding the Commonwealth. Wilson lodged a Freedom of Information request and eventually received a heavily redacted set of documents. An anonymous whistle-blower had reported him as joining the RAAF just so that they would pay for his medical transition. The allegation and investigation had been triggered in August of Wilson's first year, when only Wilson's chain of command and medical staff knew he was transgender. Wilson also lodged a complaint against the ADF in the Australian Human Rights Commission. The ADF's initial response was 'basically that everything that I had listed as happening either didn't happen or was my fault'. Wilson declined to take his case to mediation because of the strain on his mental health.[29]

The examples of both Joel Wilson and 'Mitch' reveal much about the mixed experiences of more recent transgender Defence members. There could be peers and commanding officers who were affirming. Although it would be rare for people to bully or taunt directly, the rumour mill meant a lot of harassment behind transgender people's backs. Those people who were transphobic or uncomfortable socialising with a transgender person simply avoided them, leading to a sense of isolation. These forms of harassment made it difficult for transgender members to confront discrimination directly.

Perhaps the biggest lesson from the experiences of Mitch and Joel Wilson reflects the broader experience of transgender service. When things were going well and there were no problems, which was the case for many other transgender members, they could go about their normal working lives. When there were problems, however, the ADF did not handle them well. The bureaucracy often clouded matters in secrecy and/or sanitised them, trying to disguise any underlying transphobia.

Transgender policy

As outlined in chapter 6, there was a policy void following the repeal of DI(G) Pers 16–16 in September 2010. This finally ended in April 2015 when ADF Joint Health Command released Health Directive 234: 'Medical Management of Gender Dysphoria and Gender Realignment in Defence Members'. The policy focused on medical understandings of transgender people and explained the varying medical options and mental health barriers confronting transgender people. Noting the diversity of transition journeys, the document explicitly stated that 'treatment options need to be tailored to the individual'.[30] Such an approach built upon the RAAF Diversity handbook and other education provided by transgender service members and DEFGLIS.

Other sections of Health Directive 234 contradicted this point and were prescriptive about the management of transitioning Defence members. The section on medical employment classification said that transgender members were 'generally not deployable' for anywhere between six and twelve months after they start taking hormones. It also said that gender affirmation surgery 'would generally mean a non-deployable MEC [medical employment classification] of at least six to nine months'.[31] The policy thus meant that transgender members undergoing medical transition had to be downgraded medically for long, prescribed periods, even if they were physically capable of performing their jobs. Long periods of medical downgrades can affect people's careers, as they are often interpreted as 'problems', may be denied promotions or have to show cause to continue in their roles. Donna Harding described the process under Health Directive 234 as a 'MEC merry-go-round [that] is detrimental to someone's mental health'.[32]

Health Directive 234 also posed a set of contradictory clauses around gender affirmation surgery. One point affirmed that all healthcare entitlements covered by Defence Health include 'd. Surgical procedures that meet MBS [Medical Benefits Schedule] clinical indication requirements; e. Any routine clinical care unrelated to gender dysphoria or its management, as for all other members'. The next section, however, stated: 'Procedures that will not be provided at public expense … include: a. Any gender realignment surgery … b. Hair electrolysis or removal procedures.'[33] These conflicting statements marked a huge change because they meant that transgender Defence members would potentially have to pay for their own gender affirmation surgery. The statement about gender affirmation surgery drew on mythology that such operations are cosmetic. As the World Professional Association for Transgender

Health regularly argues, gender affirmation surgery is a medically necessary, prescribed procedure.[34] Fortunately ADF doctors tended to interpret the conflicting clauses liberally to support transgender people. Transgender members had gender affirmation surgery covered by Defence Health, namely because their doctors explicitly argued that it was medically necessary. Indeed, by 2016, internal Defence documents clarified that the only operations not covered were those that did not fit Medical Benefits Schedule clinical guidelines and those that were cosmetic.[35]

Gender affirmation surgery has also become a target in the new Australian culture wars. During a June 2015 Senate Estimates Committee hearing, the independent Senator Jacquie Lambie asked the VCDF questions about payments for gender affirmation surgery.[36] The ADF provided answers a week later, and the *Daily Telegraph* headlined a story 'Sex ops high on military agenda'. The beat-up article opened: 'Taxpayers have been hit with a $648 000 bill to cover the cost of multiple sex change and breast enhancement procedures for serving members of the Australian Defence Force during the past two and a half years.'[37] In subsequent years the Murdoch press and conservative politicians continued to focus on gender affirmation surgery to attack the ADF's diversity and inclusion strategy. In October 2017, former SAS officer turned Liberal parliamentarian Andrew Hastie criticised the ADF for paying $1 million for medical treatment for transgender Defence members since 2012. Hastie asserted, 'I do not see how these surgeries enhance our war-fighting capability as a nation. It's a bad joke. Why is the ADF now a vehicle for radical social engineering?'[38]

An internal ADF briefing note indicated that from November 2012 to July 2017, the ADF provided medical treatment to thirty-two transgender Defence members, seventeen of whom had gender affirmation surgery. The $1.16 million paid amounted to .006 per cent of all expenditure on Defence health.[39] The VCDF, Vice Admiral Ray Griggs, vociferously defended the funding of gender affirmation surgery in an October 2017 Senate Estimates hearing:

> We are talking about our people. These are people who are wearing the uniform of this country and serving this country. They deserve to be treated with the respect that any other member of the Australian Defence Force is treated with. They deserve to have appropriate medical treatment that any other member of the Australian Defence Force has … The aim here is not to lose talented individuals [on] whom we have spent many, many millions of dollars in training—many millions—and many, many times more than $1 million worth of treatment. I have found the public

> debate to be almost bordering on hysterical and very, very unhelpful, and very unhelpful for our people who are managing their way through this.[40]

Defence Minister Marise Payne, too, defended the funding of gender affirmation surgery.[41] As discussed below, funding for gender affirmation surgery is only one issue for which conservatives have attacked the ADF.

In February 2019, a new policy on gender dysphoria was published in the *Defence Health Manual* to replace Health Directive 234. The new regulation addressed most of the criticisms of the earlier policy: there are no longer automatic MEC downgrades, and instead the members' MEC and other treatment are to be considered on a case-by-case basis. The new policy also clarifies that Joint Health Command will pay for the majority of medical treatment, including: 'surgical procedures that meet the MBS clinical indication requirements; laser hair removal from scrotum for male to female [MTF] to facilitate formation of a vagina; erectile device implantation'. Joint Health Command will not pay for surgery that does not meet MBS criteria, cosmetic surgeries, hair removal (other than that listed above), speech pathology for voice modulation, gamete storage (for fertility), breast augmentation, breast reduction, mastopexy (surgery to modify the size and shape of the breast) and facial feminisation procedures. To commence any medical or surgical treatment, transgender Defence members need the approval of a psychiatrist who specialises in gender incongruence.[42] This is a more rigid requirement than for transgender people in civilian Australia, who can commence hormones with just a prescription from their GPs. The ADF approach thus represents a shift away from the informed consent model touted by transgender activists, but it aligns with the ADF's cautious approach to ensure that its members are at optimal health and that treatment meets the World Professional Association for Transgender Health's standards of care.

Non-binary and intersex personnel

Non-binary and intersex members of the ADF have also had to fight hard for reform. From July 2016, under directions of the Attorney-General, all Commonwealth Government departments and agencies, including the ADF, had to update their internal systems to allow employees to identify their gender as male, female or 'Indeterminate/Intersex/Unspecified (X)'.[43] This was the first time that the ADF ever adopted a formal policy that mentioned intersex personnel. Intersex activists have argued that this classification is problematic because it falsely conflates all intersex people as non-binary, whereas the majority of intersex people identify as male or female. Internal consultations with intersex Defence members reached the same conclusion and noted: 'Changes to policy should be careful to avoid inadvertent application to intersex.'[44]

The first non-binary member to enlist and identify as 'Indeterminate/Intersex/Unspecified (X)' commenced at ADFA in 2016 as an Army aviation cadet. Riley Bradford has identified as agender since high school but is also happy to be described as trans, transgender, non-binary or trans-masculine. Bradford came out to their divisional officer shortly after they commenced at ADFA and, unlike Joel Wilson, had a smooth time through doctors' and psychiatrists' referrals. Bradford was open from the beginning that they might want some medical interventions, but they were non-binary and did not intend to transition to the opposite gender of their sex assigned at birth. Bradford found that both their chain of command and peers were generally accepting. As they summarise, 'Everyone [at] ADFA at least is just, like, "Yeah, I don't really understand how that works, or what that would feel like, but you can do your job, and you're not a bad person, so, it must be OK."' Bradford uses the pronouns they/them and prefers to correct people who misgender them. While usually it is an innocent mistake rather than malicious, still this can have an effect: 'The way I normally explain it to people is it's like getting a papercut. And one papercut you lick it and it stops bleeding and doesn't matter. But if you get a dozen papercuts every single day for years and years and years …'[45]

Like so many other pioneers, however, Bradford encountered hurdles from an institution that was highly gendered and had not yet developed any policies or procedures to support non-binary members. For instance, even though Bradford applied to update their gender marker, the ADF would not let them until they had either revised their birth certificate, updated their passport or had a doctor's letter confirming they had undergone some sort of medical treatment. Even beyond that hurdle, however, the ADF records both a member's gender *and* their sex. Although gender now has the 'unspecified' option, still the member must nominate either male or female as their sex.

Other more routine challenges related to uniform, fitness, toilets, facilities and salutations. The uniform was relatively easy: Bradford was permitted to select either male or female uniform. For fitness, Bradford was still obliged to meet the standards of their sex as recorded in the ADF system before they began medical transition. That said, they were determined to meet the male fitness standards to pre-empt criticism about their physical capabilities. Once Bradford proceeded with medical transition, their chain of command consulted with their doctor to determine that the male fitness standard was most appropriate to use, to which they have happily adhered. Toilets prove a challenge because there are only three gender-neutral, accessible toilets on the ADFA campus. On one instance when Bradford trained at the Majura Training Area, they were ordered to use the toilets corresponding to their sex assigned at birth. Bradford politely pointed out that under the Sex Discrimination Act they should be allowed to choose which toilet to use, but the instructors dismissed this. In terms of facilities, when Bradford started at ADFA all residences were segregated by gender in corridors of four people each. Bradford's divisional officer decided to mix the sections, and there were no problems or complaints within their division. Salutations have not been a problem at ADFA because all students are referred to by the gender-neutral rank and title of 'cadet'. It will be more complicated once Bradford graduates, as they would have to choose either 'Sir' or 'Ma'am'. Bradford has already come up with a possible workaround: to be called by their rank (e.g. 'Captain'). Bradford's chain of command initially rejected this proposal, but later agreed that it would be appropriate if they preferred.[46]

In 2017 another non-binary person commenced in the RAAF at ADFA to study aeronautical engineering. 'Elliott', too, had known since high school that they were non-binary but was not sure how that would be received at ADFA. Members of the ADFA LGBTI Society, including Bradford, gave a short presentation during orientation, and Elliott realised that ADFA would be an affirming environment. Elliott advised their sergeant that they were non-binary and told their division at a face-to-face meeting. Similar to Bradford, Elliott found that the 'initial sort of meeting went really well. In the couple weeks after that, it was a little bit—some people were really great about it. Some people I'd told earlier on as well. I'd told a few div members already. And pronouns are still a work in progress.' One cadet shared a Facebook video that derided Elliott, but when it was reported the cadet was reprimanded. Otherwise, the majority of cadets have consistently been supportive or at least not had a problem with Elliott.[47]

Elliott ran into similar challenges as Bradford regarding toilets, uniforms and facilities, and for Elliott the outcomes were not always as positive. For instance, they initially had to have a female uniform, but because Elliott used a binder, the shirt just looked too big. After a short argument the tailor agreed to give Elliott a male shirt, but still they had a female hat. Eventually, Elliott applied for and was approved to change fully to men's dress uniform. For facilities, whereas Bradford's divisional officer was willing to make their corridors mixed gender, Elliott's divisional officer refused to do so. As such, Elliott lived in a corridor designated only for people with Elliott's sex assigned at birth. The divisional officer also delayed issuing a new door sign in Elliott's new name, so they simply took matters into their own hands: 'I got to the point where I whited it [the sign] out and wrote my name over the top because I was, like, that's not the right name any more.'[48]

Although Bradford and Elliott's peers have accepted them and they have been patient and innovative at navigating the ADF's gendered system, still they have become unwitting pawns in conservatives' attacks on the ADF's diversity and inclusion agenda. In September 2017, the *Australian* published an article titled 'Cadet X clears a path for gender-neutral Aussie soldiers'.[49] A week later, the Minister for Defence Personnel, Dan Tehan, ordered the Department of Defence to investigate seeking an exemption from the Sex Discrimination Act provisions that had facilitated the recognition of non-binary genders. The department statement read:

> The defence of our nation is our first priority and we will examine anything that arises that may impact on our ability to achieve that. In this regard, the Government and the Australian Defence Force are currently considering the need for an exemption to the Act (*Sex Discrimination Act 1984*) and the guidelines due to the unique operational requirements of military service.[50]

Since that brief mention in the media, there has been no public statement or reports about non-binary recognition in the ADF. Still, the publicity surrounding 'Cadet X' showed just how vulnerable gender-diverse members of the ADF are to wider culture war battles.

Both Riley Bradford and 'Elliott', despite some obstacles, had relatively positive experiences at ADFA. Partly this is because they were insulated and, in many ways, there was more scope for divisional officers and the chain of command to be flexible and try new procedures. Time will tell how they will cope in the wider ADF environment when they are working with service members who are older, might have more rigid beliefs about gender binaries, and are less accommodating.

HIV and the contemporary ADF

For almost twenty years after *X v. Commonwealth of Australia*, the ADF's policies towards people living with HIV changed little despite major medical advances. A range of treatments are now available that both reduce the risk of HIV transmission and have transformed the experience of people living with HIV. Antiretroviral medications known as PrEP have significantly reduced the rate of HIV transmission by up to 99 per cent if a PrEP user adheres to the regimen of daily tablets.[51] Furthermore, HIV post-exposure prophylaxis (PEP), prescribed within seventy-two hours of exposure to HIV and continued for twenty-eight days, significantly reduces the risk of seroconverting.[52] Major studies have also shown that if someone who is HIV positive takes antiretroviral medication daily as prescribed and maintains an undetectable viral load, there is effectively no risk of them transmitting the virus.[53] Since 1996, with the advent of HAART, the experience of living with HIV has become a chronic and manageable condition.[54] Nevertheless the ADF has been slow to reflect this transformation in its policies and practices.

In May 2015, ADF Joint Health Command (JHC) received requests from two Defence members for access to PrEP, sold under the brand name of Truvada, in order to reduce the risk of HIV transmission. At the time, PrEP was being trialled in Queensland, Victoria and New South Wales, and international evidence was showing the medication was proving highly effective. The JHC Pharmacy and Therapeutics Committee agreed that PrEP, when assessed on a case-by-case basis, should be supported by the ADF. It argued that the risks of not providing the medication outweighed the potential costs and identified the risks as:

(a) The member becoming infected with HIV, which would then require lifelong antiretroviral treatment with several medications and have negative consequences for their career as they would be non-deployable.
(b) The member sourcing the medications from overseas suppliers, which may be unregulated and of unknown quality, thereby compromising safety and effectiveness.
(c) The member not notifying Defence of the treatment. Antiretroviral medications can cause serious adverse effects and individuals taking these drugs need to be appropriately monitored.[55]

In approving Truvada for two service members as PrEP, the ADF was a pioneer; indeed, it was the first major Australian organisation to do so. However, in August 2015, the *Australian* newspaper ran an article with the heading 'ADF paying $10k per person to prevent HIV in at-risk personnel'. The article asserted that 'the drug costs more than $10 000 a year per person but the ADF is covering the cost for its members, while other Australians have resorted to importing generic PrEP drugs purchased online to lower the cost'. An ADF representative told the *Australian* that PrEP was being made available only to a 'very small' number of ADF members under specialist supervision and in accordance with guidelines released by the Australian Society for HIV Medicine.[56]

By November 2015, the media reported that the ADF would no longer fund access to PrEP.[57] Under Health Bulletin No. 02/2015, Defence members would have the same options for accessing the medication as the general public. In practice, this meant that they were placed on a waiting list for 'oversubscribed' state-based trials and would have to self-fund the cost of the medication.[58] In May 2016 the Therapeutic Goods Administration finally approved Truvada for the purpose of the prevention of HIV. In April 2017, the ADF revised its policies and approved providing access to this medication for 'individuals assessed as being at ongoing risk of acquiring HIV infection'. The policy had similar guidelines to civilian sexual health clinics regarding use and regular sexual health testing.[59] Truvada was listed on the Pharmaceutical Benefits Scheme on 1 April 2018, lowering its cost substantially. In March 2019, the *Australian* reported that more than fifty ADF members had been prescribed PrEP.[60]

Although the ADF has moved to incorporate the prescription of medication that can lower the risk of HIV transmission, it has not yet altered its position on the recruitment and retention of HIV-positive personnel. In August 2015, this matter received public attention when a Federal Court judgement was handed down involving a HIV-positive RAAF airman who had been discharged from the ADF in 2014 on the grounds that he was medically unfit. The airman, known as 'C', enlisted in the RAAF in 1999 and was diagnosed with HIV in September 2011. In October 2012, following a review of the Medical Employment Classification Review Board, 'C' received a termination notice and elected to accept the termination, which was effective in January 2014. 'C' subsequently launched proceedings in the Federal Circuit Court, claiming that the Commonwealth had contravened the *Fair Work Act 2009* by dismissing him because of a disability. The Federal Court judges found that the Fair Work Act did not apply to the ADF, 'as they are not, through their enlistment, party to any contract of service'.[61] Thus the Federal Court ruling in *C v. Commonwealth of Australia* affirmed the same principle as the 1999 High Court ruling in *X v. The Commonwealth of Australia*: it was within the law for the ADF to restrict and discharge Defence members living with HIV.

Notwithstanding what happened to 'C', the policy for serving Defence members who seroconverted was not necessarily dismissal, but HIV-positive people were all downgraded to non-deployable status and were subject to ongoing medical evaluations. The experience of 'Trent', an HIV-positive gay man currently serving in the Army, shows how these policies towards HIV affect personnel. He joined the Army in 2013 and was diagnosed as HIV positive in November 2014. Upon being diagnosed, Trent informed his regimental sergeant major. He remembers, 'She was quite distraught about it, like sorry for me. She didn't react badly to it at all.' From there, they informed his immediate warrant officer and lieutenant 'because they were the ones that tasked me to go on exercises and things, so they had to know why I couldn't go on the exercises'. Trent also saw a civilian doctor, who recommended he take antiretroviral medication, 'and I've been undetectable since', meaning that HIV cannot be detected in his blood tests.[62] So long as Trent remains undetectable, the body of scientific evidence shows that he is not able to transmit HIV to others.

Although Trent's HIV status is undetectable, and despite the fact that he is one of the fittest people in his unit, he was immediately downgraded under the ADF's Medical Employment Classification (MEC) System to the category of J40: 'Holding temporary—Confirmation and allocation of suitable MEC classification pending MECRB determination'. This falls under the MEC 4 'Employment Transition' heading. His unit had recommended that he be retained in the Army, but his career adviser in Canberra, who had a personal enmity towards him, recommended discharge. At the time of our interview (2017) Trent was classified as J42, 'Employment at Service Discretion—MECRB assigned only—duration up to five years at any one time'. This means that he has 'employment at service discretion for a maximum of five years'. Trent explains that a condition under this MEC is that 'I can't be posted. So I'm in this job when I got that category, so I can't change that job.'[63] Trent's treatment by the Army, including his medical downgrade, was in line with Health Directive 210 of June 2007: all people who tested HIV positive must be downgraded to MEC 4: Employment Transition.[64]

Trent notes that his HIV status was automatically added to his medical records as soon as he disclosed his status, but 'no one is supposed to open those medical records unless they've got a reason to, unless they're putting more paperwork in'. If medical information was leaked, then the person responsible 'could just be discharged, straight away disciplinary action'. People in Trent's chain of command know that his MEC has been downgraded, but they are not allowed to ask why. Trent's commanding officer was initially 'very professional about it all, and then proceeded for the next year to basically stop me doing anything except scratching myself'. Trent states: 'We don't do field [exercises] much [in my role] so it's pretty basic, and not in any way dangerous to life or limb. But his [CO's] view was that if I was to break my leg, and someone was near me, and I was to bleed on them and they had scratched themselves that instantly they were infected.' Trent expresses frustration that under the ADF's policy on blood-borne viruses, HIV was treated differently from hepatitis: 'If you're undetectable with the heps you go up' to the deployable MEC J23, 'and you're fine—even though the hepatitis's are actually more virulent, but HIV's still got that stigma attached to it'.[65]

Trent found these limitations very frustrating, as did Trent's RSM. Trent's undetectable HIV status means that the scenario described by his CO could not occur. Trent remembers saying to his CO: '"I just want to be treated normally. I can't hurt anyone. I couldn't infect someone if I tried." And he just didn't get it. "I've got a duty of care to the rest of my members" and all this sort of thing.' Even though a new CO has been much better at understanding Trent's capabilities, his career adviser, who decides whether Trent is able to go on courses, still blocked him from taking up promotion courses. He explains, 'I was recently nearly on a big posting … and my career adviser said no, "Because they're not going to get their money out of you because it's five years maximum."' Trent points out that his MEC means that his medical condition is reviewed every five years and had been extended at each of the two reviews he has had so far.[66]

Trent has refrained from telling many people in his life about his HIV status and from seeking support from such organisations as the state AIDS councils. The ADF is a closely knit world, and he fears other service personnel finding out. He does wish that he could tell his friends within the ADF about his status, but he worries that news about his status would spread. When asked what he fears would happen if people found out, Trent says:

> That they wouldn't want to be in the same room as me. I mean I've got incredible, incredible feelings of guilt over getting [HIV]. Yeah, I burst into tears sometimes. And incredible feelings of guilt of the disappointment, of course, my partner, of how I've just fucked my career. There's so many things that I was wanting to do this time and I've just kicked it in the nuts, and I feel bad trying to push to do these things, and push policy and everything, because I think, 'Well, it's your own fucking fault. You shouldn't have fucked up, or you should have actioned it as soon as it did happen. You should have done something about it when you could have.' And it's those feelings of guilt and everyone keeps saying, 'You shouldn't feel guilty. It's happened, it's happened. You got to live with it. It'll just …' But you can't help feeling that guilt. And then I think to myself, 'Well, I'm not special. There's not many of us in the military. Why should they go to the trouble and expense of changing policy just for a couple of people?' But the trouble is there's every chance there's going to be a lot more people.

Trent's civilian doctor informed him that she knew another man serving in the ADF who was HIV positive. Trent asked her to pass on his contact details, and the two have since spoken. The other man has been in the ADF for almost three decades and was diagnosed at a similar time as Trent. Trent is also aware of other HIV-positive people who were forced out of the ADF because of their status: 'So in other words, thank heavens for small mercies. I'm still in; I'm well respected in my job. I may not be able to do as much as the other people, but it's getting better.'[67]

Our interview with Trent was in 2017; in 2018 the ADF quietly updated its policy on HIV and other blood-borne viruses. Although HIV-positive people are still not allowed to enlist, for the first time there is the possibility of deployment for serving HIV-positive Defence members. The *Defence Health Manual* chapter on blood-borne viruses says that the Medical Classification Review Board may assign a deployable MEC (J29, L27, L28, M25) if the member has:

(a) sustained undetectable viral load
(b) normal immune function
(c) no adverse effects from their treatment
(d) no complications from their infection
(e) treatment and monitoring requirements that can be managed in the deployed environment.[68]

This change came about after the Surgeon General of the Australian Defence Force, Air Vice Marshal Tracy Smart, worked closely with stakeholders and the VCDF, Ray Griggs. They wanted to ensure that the ADF's policies on HIV—including access to PrEP—were up to date with the science and societal expectations. Griggs explains: 'It's compelling from my perspective; if you've got low viral loads and all that sort of stuff, and you've got a range of treatment options post-exposure. For me, it was a bit like Dark Ages stuff; we'd held on to Grim Reaper and we hadn't moved at all.'[69] Griggs also ensured that the ADF had a public relations strategy in place, anticipating that this policy could be another culture war trigger point. At the time this book was published, this has yet to happen.

The changed policy has already had a positive effect on Trent's career: his MEC was upgraded to L27 in 2018, and he has since been allowed to do courses, participate in field exercises, deploy and has even been promoted. In an email he writes: 'Apart from doing CMEC reviews every couple of years and having to get a waiver to deploy, I'm no different to anyone else …'[70]

The role of leadership

Discrimination still affects LGBTI Defence members. Some of this prejudice has come from within the ADF, but a significant amount has come from external forces such as the conservative media. Leadership has been particularly important at advancing LGBTI inclusion within the ADF and to combat homo/bi/trans/interphobia. Such leadership has come from high-ranking

ADF officials such as David Morrison, who was Chief of Army from June 2011 to June 2015, and Ray Griggs, who was Chief of Navy from June 2011 to June 2014, then VCDF from June 2014 to June 2018. For instance, Griggs set up the Navy Diversity Forum in 2014, appointing strategic diversity advisers to the Chief of Navy, including Stuart O'Brien as LGBTI adviser.[71] Critical leadership has also come from DEFGLIS, which has worked hard to address the needs of LGBTI service personnel and advocate for full equality within the ADF.

Conservative media have taken numerous opportunities to attack the ADF's support for LGBTI inclusion along with a raft of other equity groups such as Muslims and women. Regularly these critics say that the ADF is embarking on a 'social engineering' project. As mentioned above, transgender and non-binary service have been prominent targets and, as discussed below, the ADF's participation in the Sydney Gay and Lesbian Mardi Gras parade is another perpetual target. But even smaller initiatives have met the ire of conservative media and online trolls. For instance, in 2013 Morrison authorised the Australian Army to issue a rainbow flag lapel pin and cufflinks, which soldiers may wear the week before Mardi Gras.[72] Conservative *Daily Telegraph* columnist Miranda Devine described the design as 'a blatantly political symbol, disrespecting the Anzac spirit in order to make a contentious political point, in direct contravention of official Army guidelines in place since 1903 to protect Australian Army emblems'.[73] In early 2018, a Department of Defence employee set up a rainbow flag icon that LGBTI allies could mark next to their names in the Defence Corporate Directory. Again, the *Daily Telegraph* and other conservative media attacked the ADF, alleging that this was a form of bullying against those Defence members who did not identify as allies.[74] The ADF quickly suspended this plan because it had not been approved through proper channels. Finally, in 2017 members of the ADFA LGBTI Student Group—formed around 2013 after a gay student suicided—prepared their own ADFA LGBTI Staff Guide. The document has comprehensive information about all things LGBTI, including the importance of asking people their preferred pronouns and using gender-neutral language when referring in the abstract to relationships or gender identities.[75] In August 2018, the Murdoch press falsely reported that the ADF was banning its members from using gendered pronouns, using the ADFA LGBTI Staff Guide as its 'evidence'.[76] Even the ABC program *Media Watch* condemned the *Daily Telegraph* and right-wing radio commentators for their inaccurate reports.[77]

Top-level ADF staff have consistently dismissed claims about 'social engineering' and defended the organisation's approach to diversity and inclusion. The 'Defence Diversity and Inclusion Strategy, 2012–2017' states:

> Diversity and Inclusion in Defence is a critical capability issue. The Defence organisation of the 21st Century must harness the broadest talents if we are to remain fully ready to defend Australia. In the competitive labour market for talent in Australia and with a globalized workplace Defence can no longer rely on a workforce drawn from a narrow pool of talent.[78]

A range of other ADF policy documents support inclusive policies towards LGBTI members.[79] In 2017, VCDF Ray Griggs outlined why the ADF valued diversity and was moving to reflect contemporary society:

> We are not undertaking a social experiment; we are a war-fighting organisation, one that has been on continuous operations in multiple theatres since 1990. Maintaining our ability to fight and win is paramount. At the end of the day we are trying to build a better ADF, not only technologically but culturally and behaviourally. We cannot do that by remaining anchored in a cultural context of past decades, which reflected vastly different societal views and norms to that of the society we are part of today.[80]

Griggs was consistently one of the most vocal champions of LGBTI inclusion from the top brass, and he had the full support of the other chiefs. In April 2016, the five chiefs jointly signed a letter to the *Australian* in response to accusations of social engineering: 'Diversity is not about identity politics[;] it is about … gaining a wider range of perspectives to make better decisions and, in the military context, enhancing our capability.'[81]

DEFGLIS has also shown critical leadership, providing a continuous sense of community and support for LGBTI service personnel and raising the visibility of LGBTI military service. When DEFGLIS was founded in 2002 as the Defence Gay and Lesbian Information Service, its primary advocacy was for same-sex partner recognition. In 2011, in order to be more inclusive towards transgender and intersex personnel, it rebranded as the Defence LGBTI Information Service. In 2012, on a revamped website, Stuart O'Brien asserted: 'Many people within DEFGLIS felt their job was done once anti-discriminatory policies were in place and benefits had been extended to same-sex partners; however, several Defence incidents since 2011 have caused DEFGLIS to review whether the organization is doing enough to support LGBTI personnel within Defence.'[82]

From 2012 to 2018, Vince Chong served as president of DEFGLIS.[83] In 2013, he was awarded an Australian Defence Force commendation for services delivered to improve diversity and inclusion. In March 2017, Chong and transgender RAAF Squadron Leader Catherine Humphries were recognised by *Cosmopolitan* magazine as among Australia's most influential LGBTI people because of the roles they had undertaken to support LGBTI Defence members.[84]

Sydney Gay and Lesbian Mardi Gras

One of the most high-profile ways that ADF leadership and DEFGLIS have worked together to show support for LGBTI service personnel was by enabling them to march in Mardi Gras in uniform.[85] DEFGLIS had been marching in civilian attire since 2008, and after years of lobbying behind the scenes, the CDF approved for them to march in uniform in 2013. The ADF public statement said, 'Diversity is a strength and asset for today's employers and Defence is no exception. Workplace inclusion for all ADF members is a high priority for the organisation as it undergoes cultural change.'[86] The *Canberra Times* declared this as 'one small uniformed step for military personnel but a giant leap for gay and lesbian recognition'.[87] The Defence contingent, which consisted of 120 people, was led by then Air Commodore (later Air Vice Marshal) Tracy Smart, the ADF's highest ranking openly LGBTI person.

Participants included an engaged lesbian couple, Corporal Renae Fritzell-Flint and Corporal Danielle Gurkin, both physical trainers from the Army Recruit Training Centre at Kapooka. Fritzell-Flint had marched in Mardi Gras since 2001 but emphasised how special it was to be able to march in uniform alongside her partner: 'To be able to wear military uniform, something we do every day, is great. We wear it with pride and to be recognised after 20 years of being discriminated against is fine. There is a very open and accepting cultural diversity in the Army. We work together in the same unit and same section and our bosses are excellent.'[88] RAAF member Neal Fischer remembers of that first march: 'When we marched in uniform that year [there was a] massive, *massive* reaction from the crowd, but you're still military so you still march in a straight line, you got your arm going high and you're kind of looking sideways [with a] massive big grin on your face. But yeah it was awesome, an awesome feeling.'[89]

As with previous years when Defence members had marched in Mardi Gras, there was resistance to DEFGLIS marching. The visibility of ADF uniforms attracted particular ire with letters to newspaper editors showing divided opinion on whether marching in uniform was appropriate.[90] Opposition also came from former and serving ADF members who invoked their disapproval on the grounds that Mardi Gras was a political event. Although the Mardi Gras abandoned its overt political messages in 1981, Dennis Altman argues: 'In a society which is still ambivalent about homosexuality, which is far from accepting it as just another form of human sexuality and emotion, the very visibility of Mardi Gras is itself a political statement.'[91] Defence instructions relating to 'Political activities of Defence personnel' explicitly prohibit individual Defence members from wearing their uniforms when engaging in political activities.[92] One Vietnam veteran complained to the CDF (now Governor-General), General David Hurley: 'The uniform should not be used as a tool by any group, minority or otherwise, to bring attention to their cause.'[93]

The most vocal complaints came from Bernard Gaynor, who argued that the Mardi Gras was not only political but also promoting a lifestyle that was un-Christian. He maintained that by permitting LGBTI members to march in uniform, the ADF was violating its rules about engaging in unacceptable behaviour and offending his rights as a Christian. Gaynor's formal complaint denounced what he viewed as the highly sexualised nature of Mardi Gras.[94] Defence ultimately determined that while the Mardi Gras itself was not political *per se*, the ADF needed to tread carefully so that its members were not seen to be participating in political debates (i.e. same-sex marriage) that other Mardi Gras floats were advocating. This, of course, has not stopped conservatives from attacking the ADF participation in the Sydney Gay and Lesbian Mardi Gras as part of their culture war.[95]

DEFGLIS continued to organise Defence members to march in uniform after 2013. DEFGLIS President Vince Chong stated in 2014: 'I think it's great to see the future generation of ADF leaders understanding the change happening in the wider Defence Force and walking the talk—embracing inclusion right from the outset of their initial officer training. They recognise that inclusion is the key to greater productivity, cohesion and results from human capital.'[96] Several interview participants have marched in Mardi Gras, with some sample comments describing it as 'Oh, liberating. Loved it' ('Pete', Navy); 'Oh, it was awesome. So—but I mean, I was absolutely terrified' (Loretta Cincotta, RAAF); 'It was really good. I thought I'd be nervous, [but I] wasn't. I think once we started, I just kind of tuned out and just concentrated on what I was doing, like, [I] didn't want to fuck up because the whole world was there' (Russ Tolland, Army).[97] Media reports in 2015 suggested that the ADF and the NSW Police received the loudest cheers that year from parade viewers.[98] In 2015, the Army, Navy and Air Force, most senior enlisted officers volunteered to lead the Defence contingent. Air Force Warrant Officer Mark Pentreath said: 'Why wouldn't I be proud [to lead the contingent]? These men and women are part of the team that is our future as an ADF. To me, marching in the Mardi Gras parade is no different to representing the Air Force at any cultural event that is important to our people such as White Ribbon Day, or International Women's Day.'[99] The ADF has continued its participation in Mardi Gras every year.

Other celebrations of LGBTI service

In September 2015, DEFGLIS hosted the first annual Military Pride Ball in Sydney. The *Star Observer* described the 2015 Military Pride Ball as the largest gathering of LGBTI service personnel since the ban on LGB service was lifted in 1992.[100] A new Defence Pride Network was also officially launched at the event, in line with existing women's networks and a disability employment network in place within the ADF.[101] DEFGLIS also presented its first annual Awards for Excellence, recognising the contributions that Defence members, public servants and allies have made to support LGBTI inclusion.

Over the years the Military Pride Ball has attracted sponsorship from the private sector and hosted several hundred guests from across the ADF, LGBTI community, ally organisations and veterans. Guest keynotes have been human rights commissioners, the VCDF, the Surgeon General of the Australian Defence Force, the Chief of Joint Operations and the Deputy Chief of Army. DEFGLIS President and then RAAF Squadron Leader (now Group Captain) Vince Chong described the Military Pride Ball as a milestone on the ADF's journey of acceptance for LGBTI staff: 'The Defence organisation over the past 20 years has moved from conducting witch hunts to ferret out "the gays" to encouraging a culture of respect and inclusion.' He also stated that one of the biggest areas of progress was the increasing level of support and acceptance of transgender people who have been allowed to serve since 2010. Chong said: 'Three to five years ago there as an assumption was that transgender people should just … be invisible. That is no longer the case, but we still have work to do.'[102]

DEFGLIS operates as a social, advocacy and peer support group, so during the year its board plans a range of other, smaller activities—as does the ADFA LGBTI Student Group. These include organising a ski trip, Wear It Purple Day breakfasts, IDAHOBIT morning teas and other smaller social gatherings. In 2015, DEFGLIS turned its attention to the sacrifice and contribution of past generations of LGBTI military service by organising commemorative events on Anzac Day. DEFGLIS arranged for current and former LGBTI servicemen and women to lay rainbow-coloured wreaths at the Shrine of Remembrance in Melbourne, the Australian War Memorial in Canberra, the Cenotaph at Martin Place in Sydney and in Townsville. In Melbourne, Max Campbell—a former member of the Gay Ex-Services Association—was chosen to lay the wreath, flanked by currently serving members of the Army, Air Force and Navy.[103] DEFGLIS continues to organise rainbow wreath-layings on Anzac Day and has facilitated them in Perth, Brisbane, Adelaide and Daylesford, Vic. A post on the DEFGLIS website explains:

> DEFGLIS participates in Anzac Day because this day is important to all Australians. It is a day where we can celebrate our shared values as Australians and be proud of who we are. Wreath-laying is an activity that seeks to recognise all who served. We do not know who they all were, but they don't deserve to be forgotten. The rainbow wreaths placed by DEFGLIS incorporate respectful commemoration of LGBTI personnel who served, and recognition about the effects that the wars had on their families.[104]

The rainbow wreath-layings thus bring the history of LGBTI military service full circle: they join the past and the present and are a reminder that the openness and benefits LGBTI Defence members enjoy today came after more than a century of secrecy and sacrifice from their forebears.

Conclusion

Since 2008, the ADF has moved to a position where it actively recruits LGBTI service personnel, recognising the diversity and capabilities they provide to the organisation. There has been some resistance to this more inclusive approach, but the support of high-ranking officers and the advocacy of DEFGLIS have helped to advance LGBTI equality. Defence Force Recruiting representatives regularly attend LGBTI events, such as Mardi Gras Fair Day, Melbourne's Midsumma Carnival, Canberra's SpringOUT Fair Day and Adelaide's Feast Festival. Defence Force Recruiting has also advertised in LGBTI media, including *LOTL* and *DNA* magazines. LGBTI inclusion has been incorporated in ADF strategic documents, and guides have been prepared to provide information about LGBTI-inclusive policies. In the most recent Defence Census 2015 Public Report, 3.9 per cent of respondents from the permanent forces identified as LGBTI; this broke down to 5.1 per cent of Navy, 3.2 per cent of Army and 4.1 per cent of RAAF respondents.[105]

The changes that the ADF has made are substantial, but there is still a distance to be traversed and progress cannot be taken for granted. Conservative media and politicians have often attacked the ADF for its inclusion policies, arguing that it is engaging in social engineering and detracting from its core mission to defend Australia. These attacks have especially targeted transgender and gender-diverse members, who just want to go about their lives in peace and be recognised in their affirmed genders. It is not surprising that conservatives target the ADF, given the significance of the Anzac mythology to Australians' sense of national identity. Moreover, given the restrictions on Defence members making statements to the media, there is little that LGBTI service members can do to respond to such attacks. Although the ADF leadership has consistently defended the inclusion agenda, they also must follow the orders of the government of the day. If there is a conservative government, then speaking up too loudly runs the risk of the government imposing new policies or practices that do not support LGBTI service members.

It is perhaps because of this wider culture war climate that the ADF has recently gone quiet on its inclusion agenda. For instance, the ADF has yet to publish an updated Diversity and Inclusion Strategy on its website, the 2012–17 strategy still being posted. The RAAF handbooks on transitioning gender and on sexuality are no longer publicly available on the website and instead are only available internally to Defence members. The ADF backed away from the ally initiative for the Defence Corporate Directory, and there are fewer media releases and announcements promoting LGBTI events besides Mardi Gras and the Military Pride Ball.

Recent public debates about LGBTI inclusion nationally and internationally show that the victories of LGBTI rights and recognition in the ADF cannot be taken for granted. In the United States in 2017, for example, President Donald Trump tweeted: 'The United States Government will not accept or allow Transgender individuals to serve in any capacity in the U.S. Military.' This marked a reversal of reforms commenced under President Barack Obama in 2016 to open the US armed forces to transgender members. Since 12 April 2019, transgender military personnel have not been allowed to serve or enlist in the US military, except if they serve in their sex assigned at birth, had already transitioned or were given a waiver.[106]

The American circumstances, of course, are quite different, as the ban on LGB service was lifted only in 2011, and the lifting of the transgender ban was never fully implemented. But the lesson to take from the United States is that advocates for LGBTI inclusion cannot take the status quo for granted. Conservative attacks mean that LGBTI Defence members and their allies must be vigilant and visible to defend the importance of inclusion in the ADF. As history has shown, LGBTI Defence members and their allies are well and truly a part of the ADF landscape, and there is every reason to be optimistic that there will always be another generation ready to fight for their rights.

Notes

1 Cosima Marriner, 'Twenty years on, great strides for full inclusion in the military', *Sydney Morning Herald*, 13 May 2012, p. 8.

2 Department of Defence, 'Defence Diversity and Inclusion Strategy, 2012–2017', https://www.defence.gov.au/Diversity/_Master/docs/DDIS-12-17.pdf, retrieved 11 January 2020.

3 Captain Dominic Lopez, 'Brothers and sisters in arms: Experiences of gay soldiers in the Australian Army', *Australian Army Journal* 10, no. 3 (2013): 215; Ian McPhedran,

'Gay soldiers safe from harassment but hide sexuality survey finds', 9 September 2013, https://www.news.com.au/national/gay-soldiers-safe-from-harassment-but-hide-sexuality-survey-finds/news-story/fd40b4ae74f3409ebe33a332675ff45f, retrieved 13 January 2020.

4 Scott Amos, interview with Noah Riseman, 28 August 2017, Townsville.
5 'Linda', interview with Noah Riseman, 31 July 2016, Canberra.
6 Connor Haas, interview with Noah Riseman, 20 June 2017, Canberra.
7 Graeme Billett, interview with Noah Riseman, 17 May 2018, Canberra.
8 Neal Fischer, interview with Graham Willett, 13 September 2016, Ipswich, Qld.
9 Rob Brown, interview with Noah Riseman, 8 July 2017, Sydney.
10 Becky Green, interview with Noah Riseman, 21 September 2017, Sydney.
11 'Diego', interview with Noah Riseman, August 2017.
12 'Larissa', interview with Noah Riseman, 25 August 2017, Brisbane.
13 'Cooper', interview with Noah Riseman, 23 September 2016, Sydney.
14 'Lexa', interview with Noah Riseman, 19 September 2016, Canberra.
15 'Larissa', interview with Noah Riseman, 25 August 2017, Brisbane.
16 'Lexa', interview.
17 Adam Pritchard, interview with Noah Riseman, 2 September 2016, Melbourne.
18 James Massola, 'Defence still investigating anti-gay Facebook page', *Australian*, 13 April 2011, https://www.theaustralian.com.au/national-affairs/defence-is-still-investigating-an-anti-gay-facebook-page-unwilling-to-say-whether-perpetrators-punished/news-story/70c35930829f8597d1df2c118bf28022, retrieved 13 January 2020.
19 Nick McKenzie, 'Gay soldiers blast "savage" online hate campaign', *Sydney Morning Herald*, 14 April 2011, https://www.smh.com.au/technology/gay-soldiers-blast-savage-online-hate-campaign-20110413–1de7j.html, retrieved 13 January 2020.
20 'NSW: Creator of homosexual "hate" page charged', *AAP General News Wire*, 5 May 2011.
21 *Gaynor v Chief of the Defence Force (No 3)* [2015] FCA 1370; 237 FCR 188.
22 *Chief of the Defence Force v Gaynor* [2017] FCAFC 41.
23 Catherine Humphries, interview.
24 Liam Bone, interview with Noah Riseman, 3 July 2017, Newcastle.
25 Joel Wilson, interview.
26 'Mitch', interview.
27 Joel Wilson, interview.
28 Joel Wilson, interview.
29 Joel Wilson, interview.
30 Department of Defence, Health Directive 234, 'Medical management of gender dysphoria and gender realignment in Defence members', 13 April 2015.
31 Department of Defence, Health Directive 234, points 29 and 33.
32 Donna Harding, interview.
33 Department of Defence, Health Directive 234.
34 World Professional Association for Transgender Health, 'Position statement on medical necessity of treatment, sex reassignment, and insurance coverage in the USA', 21 December 2016, https://www.wpath.org/newsroom/medical-necessity-statement, retrieved 9 January 2020.
35 Air Vice Marshal Tracey Smart, to Assistant Minister for Defence, 20 January 2016, Defence FOI 009/17/18.
36 Senate, Foreign Affairs, Defence and Trade Legislation Committee, Official Committee Hansard, 1 June 2015, pp. 42–3.
37 Ian McPhedran, 'Sex ops high on military agenda', *Daily Telegraph*, 9 June 2015, p. 22.
38 Greg Brown, 'Hastie blasts taxpaper-funded sex-change surgery for ADF', *Weekend*

Australian, 14 October 2017, https://www.theaustralian.com.au/nation/defence/hastie-blasts-taxpaperfunded-sexchange-surgery-for-adf/news-story/b26cc28a1f90ab9f136991b84966d2c7, retrieved 9 January 2020.

39 'Senate Estimates brief: Personnel 26a medical management of gender dysphoria', 18 October 2017, Defence FOI 001/18/19.

40 Vice Admiral Ray Griggs, Senate, Foreign Affairs, Defence and Trade Legislation Committee, Official Committee Hansard, 25 October 2017, p. 104.

41 Paul Karp and Joshua Robertson, 'Marise Payne defends ADF provision of gender reassignment surgery', *Guardian*, 16 October 2017, https://www.theguardian.com/australia-news/2017/oct/16/marise-payne-defends-adf-provision-of-gender-reassignment-surgery, retrieved 9 January 2020.

42 *Defence Health Manual Vol 02 Part 09*, chapter 13, 'Gender dysphoria'.

43 DEFGRAM No 297/2016, 'Recognition of gender in Defence—Gender data collection and reporting for Australian Defence Force and Australian Public Service Employees', 7 July 2016.

44 Email from [redacted] to Colonel Philip Hoglin, 'Consultation with intersex and non-binary gender service personnel', 17 October 2017, Defence FOI 001/18/19.

45 Riley Bradford, interview with Noah Riseman, 20 June 2017, Canberra.

46 Riley Bradford, interview. When reviewing this text, Bradford indicated that since the time of their interview (2017), they had witnessed a lot of positive changes regarding the specific day-to-day challenges.

47 'Elliott', interview with Noah Riseman, 20 June 2017, Canberra.

48 'Elliott', interview.

49 Joe Kelly, 'Cadet X clears a path for gender-neutral Aussie soldiers', *Australian*, 22 September 2017, https://www.theaustralian.com.au/nation/defence/cadet-x-clears-a-path-for-genderneutral-aussie-soldiers/news-story/be3c65f590aab3b1e8e9c02cdddfd108, retrieved 9 September 2020.

50 Harley Dennett, 'Defence walks away from X-inclusive gender diversity', *Mandarin*, 22 September 2017, https://www.themandarin.com.au/83956-defence-walks-away-x-inclusive-gender-diversity/, retrieved 9 January 2020.

51 Vincent J. Cornelisse, Christopher K. Fairley, Mark Stoove et al., 'Evaluation of Preexposure (PrEP) eligibility criteria, using sexually transmissible infections as markers of Human Immunodeficiency (HIV) risk at enrollment in PrEPX, a large Australian HIV PrEP trial', *Clinical Infectious Disease* 67, no. 12 (2018): 1847.

52 James O. Kahn, Jeffrey N. Martin, Michelle E. Roland et al., 'Feasibility of Postexposure Prophylaxis (PEP) against Human Immunodeficiency viral infection after sexual or injection drug use exposure: The San Francisco PEP study', *Journal of Infectious Diseases* 183 (2001): 707.

53 Robert W. Eisinger, Carl W. Diettenbach and Anthony S. Fauci, 'HIV viral load and transmissibility of HIV infection', *Journal of the American Medical Association* 321, no. 5 (2019): 451–2.

54 See also Ware, *HIV Survivors in Sydney*.

55 'Ministerial advice: Cessation of funding for pre-exposure prophylaxis for HIV infection—Truvada', 31 August 2015, Defence FOI 212/15/16.

56 Sean Parnell, 'ADF paying $10 000 per person to prevent HIV in at-risk personnel', *Australian*, 6 August 2015, https://www.theaustralian.com.au/nation/defence/adf-paying-10k-per-person-to-prevent-hiv-in-atrisk-personnel/news-story/4e971ccb2a2078a35c7c84846f0937dc, retrieved 11 January 2020.

57 Drew Sheldrick, 'Defence to stop funding HIV pill', SBS, 4 November 2015, https://www.sbs.com.au/programs/article/2015/11/03/defence-stop-funding-hiv-prevention-pill, retrieved 11 January 2020.

58 'Ministerial advice: Cessation of funding for pre-exposure prophylaxis for HIV infection—Truvada', 31 August 2015, Defence FOI 212/15/16.
59 Surgeon General Australian Defence Force, 'HIV pre-exposure prophylaxis for ADF members', Health Bulletin No. 01/2017, 18 April 2017.
60 Sean Parnell, 'More Australians taking anti-HIV drug', *Australian*, 25 March 2019, p. 10.
61 *C v. Commonwealth of Australia* [2015] FCAFC 113; Tessa Akerman, 'RAAF sacked serviceman for having HIV, Federal Court backs move', *Australian*, 27 August 2015, https://www.theaustralian.com.au/nation/defence/raaf-sacked-serviceman-for-having-hiv-federal-court-backs-move/news-story/5e2690c0ceaabc7186ba0ad14ff56a9d, retrieved 11 January 2020.
62 'Trent', interview with Noah Riseman, August 2017.
63 'Trent', interview.
64 'Health Direction No. 210: Policy on the Prevention, Detection and Management of Blood Borne Virus Infection', 13 June 2007.
65 'Trent', interview.
66 'Trent', interview.
67 'Trent', interview.
68 'Prevention, detection and management of blood-borne infection', *Defence Health Manual*, Vol. 02 Part 08, 2018.
69 Ray Griggs, interview.
70 'Trent', email to Noah Riseman, 11 June 2020.
71 Pup Elliott, 'Navy chief says diversity is essential to optimal capability', *Navy Daily*, 4 June 2014, http://news.navy.gov.au/en/Jun2014/Fleet/1102/Navy-Chief-says-diversity-is-essential-to-optimal-capability.htm#.WWMMOE0UmUk, retrieved 7 September 2017; Ray Griggs, interview; Stuart O'Brien, interview.
72 Nick Butterly, 'Army defends gay pride pins', *West Australian*, 4 December 2013, https://thewest.com.au/news/australia/army-defends-gay-pride-pins-ng-ya-362079, retrieved 13 January 2020.
73 Miranda Devine, 'The Army should be non-political. So why is it at Mardi Gras?', *Daily Telegraph*, 5 March 2017, https://www.dailytelegraph.com.au/rendezview/the-army-should-be-nonpolitical-so-why-is-it-at-mardi-gras/news-story/57693da79c45f2e75073a9cf9b2a19ba, retrieved 13 January 2020.
74 Matthew Bens, 'Defence call for public support from "allies" of LGBTI soldiers', *Daily Telegraph*, 14 February 2018, https://www.dailytelegraph.com.au/news/nsw/defence-call-for-public-support-from-allies-of-lgbti-soldiers/news-story/31d5490bf8beff3c9518aeb7a61680be, retrieved 13 January 2020.
75 'ADFA Staff LGBTI Guide 2017.' See also Riley Bradford, interview; Connor Haas, interview.
76 Ben Graham, 'Australian Defence Force denies it has banned employees using "he" and "she"', 6 August 2018, https://www.news.com.au/finance/work/at-work/australian-defence-force-denies-it-has-banned-employees-using-he-and-she/news-story/125c5410498ad83a52407edc0b087f63, retrieved 13 January 2020; Christopher Dore, 'Army drops the H-bomb: Soldiers told words like "him" or "her" can be gender bullying', *Daily Telegraph*, 6 August 2018, p. 1, and https://www.dailytelegraph.com.au/news/army-drops-the-h-bomb-soldiers-told-words-like-him-or-her-can-be-gender-bullying/news-story/dbaf4f3e2cec8912fd695fefade3a840, retrieved 13 January 2020.
77 'The Tele's gender wars', *Media Watch*, 20 August 2018, https://www.abc.net.au/mediawatch/episodes/the-tele's-gender-wars/10143828, retrieved 13 January 2020.
78 Department of Defence, 'Defence Diversity and Inclusion Strategy, 2012–2017', p. 4.

79 See for example DI(G) PERS 50–1, 'Equity and Diversity in the Australian Defence Force'; 'Navy Diversity and Inclusion Strategy 2014–2019'; 'The Army LGBTI Strategy 2016–2020'; the *Air Force Transitioning Gender Guide* and the *Air Force Diversity Guide: For Lesbian, Gay, Bisexual Members*, September 2012.
80 Ray Griggs, 'Order of Australia ADF Oration', 2017, courtesy Ray Griggs.
81 Letter from the five service chiefs to *Australian* (newspaper), 11 April 2016, Defence FOI 369/15/16. An edited version was published as a 'Letter to the Editor', *Australian*, 13 April 2016, p. 13.
82 'DEFGLIS relaunches for 2012', http://pandora.nla.gov.au/pan/132273/20120321–1646/www.defglis.com/index.php/en/media-releases/68-launch-of-new-website-2.html, retrieved 11 January 2020.
83 Riseman, Robinson and Willett, *Serving in Silence?*, pp. 241–67.
84 'Vince Chong' and 'Catherine Humphries', *Cosmopolitan*, March 2017.
85 Riseman, '"Just another start to the denigration of Anzac Day"', pp. 35–51.
86 Natalie O'Brien, 'ADF gets in uniform for Mardi Gras', *Sunday Age*, 23 December 2012, p. 6.
87 Tim Barlass and Fleta Page, 'Defence in step with gay respect', *Canberra Times*, March 2013, p. 8.
88 James Gorman, 'Marchers win Mardi Gras battle', *Central Magazine*, 27 February 2013, p. 7.
89 Neal Fischer, interview.
90 'Soldiers join Mardi Gras ranks in dress uniform', *Sunday Telegraph* (Sydney), 30 December 2012, State section, p. 78.
91 Dennis Altman, 'It began by accident', in *Mardi Gras! True Stories*, ed. Richard Wherrett (Ringwood, Vic: Viking, 1999), p. 9; Ian Marsh and Larry Galbraith, 'The political impact of the Sydney Gay and Lesbian Mardi Gras', *Australian Journal of Political Science* 30, no. 2 (1995): 308–9.
92 Defence Instructions (General) PERS 21–1, AMDT no. 3, 'Political activities of Defence personnel', 4 October 2007.
93 Letter to General David Hurley, 17 April 2013.
94 Australian Army, Defence Command Support Training Centre, Defence Intelligence Training Centre, 'Harassment brief—Maj B. Gaynor against CDF', 17 April 2013.
95 Devine, 'The Army should be non-political'.
96 Serkan Ozturk, 'Cadets to march with DEFGLIS at Mardi Gras', *Sydney Star Observer*, 28 February 2014, https://www.starobserver.com.au/artsentertainment/festival-guide/mardi-gras/cadets-to-march-with-defglis-at-mardi-gras/119263, retrieved 11 January 2020.
97 'Pete', interview with Noah Riseman, 18 July 2018, Melbourne; Loretta Cincotta, interview with Noah Riseman, 1 August 2016, Canberra; Russ Tolland, interview with Noah Riseman, 24 August 2017, Brisbane.
98 Luke Malpass, 'Sydney Gay and Lesbian Mardi Gras 2015 floods the streets of Sydney with colour', *Sydney Morning Herald*, 8 March 2015, http://www.smh.com.au/nsw/sydney-gay-and-lesbian-mardi-gras-2015-floods-the-streets-of-sydney-with-colour-20150307–13xyzi.html, retrieved 2 July 2015.
99 'Service warrant officers to march at Mardi Gras 2015', 11 February 2015, https://www.defglis.com.au/index.php/news/168-service-warrant-officers-to-march-at-mardi-gras-2015, retrieved 2 July 2015. See also 'Senior Defence Force members to lead ADF contingent in Mardi Gras parade', *RN Breakfast*.
100 Benedict Brook, 'I'm proud of the contributions LGBTI people make to Defence: Senior military chief', *Star Observer*, 8 September 2015, https://www.starobserver.com.au/news/national-news/im-proud-of-the-contributions-lgbti-people-make-

to-defence-senior-australian-military-chief/140530, retrieved 11 January 2020.

101 'Defence pride network launched', Australian Pride Network, 15 September 2015, http://australianpridenetwork.com.au/defence-pride-network-launched/, retrieved 13 January 2020.

102 Brook, 'I'm proud of the contributions LGBTI people make to Defence'.

103 Noah Riseman, 'Laying wreaths for Australians who once served in silence', *Conversation*, 23 April 2015, http://theconversation.com/laying-wreaths-for-australians-who-once-served-in-silence-40570, retrieved 2 July 2015; Benedict Brook, 'Hopes first official commemoration of gay diggers will start "new tradition" on ANZAC Day', *Star Observer*, 23 April 2015, http://www.starobserver.com.au/news/local-news/hopes-first-official-commemoration-of-gay-diggers-will-start-new-tradition-on-anzac-day/135370, retrieved 2 July 2015; 'DEFGLIS to commemorate LGBTI service', *Gay News Network*, 10 April 2015, http://gaynewsnetwork.com.au/news/national/defglis-to-commemorate-lgbti-service-17250.html, retrieved 2 July 2015.

104 Vince Chong, 'Honouring all our Diggers: Why we lay rainbow wreaths', 2017, https://www.defglis.com.au/index.php/news/267-honouring-all-our-diggers-why-we-lay-rainbow-wreaths, retrieved 13 January 2020.

105 'Defence Census 2015 Public Report', April 2016, p. 10.

106 Susan Gluck Mezey, *Transgender Rights: From Obama to Trump* (New York: Routledge, 2019).

Conclusion

Officially, LGB people were banned from serving in the ADF until November 1992. The ban on transgender service lasted even longer, until September 2010. Intersex people have had to fight for visibility, with the silence surrounding their service only now being broken. Yet, while the ADF might have tried to keep LGBTI people out of the military, they have always been there. For most of Australia's military history, they were forced to hide who they were, facing discharge if their sexuality or gender identity was exposed. The ADF leadership knows this and has probably always known it. The former Chief of Army, Lieutenant General David Morrison, acknowledges:

> There have been at least thousands of gay and lesbian men and women who have served in our Army since its inception the 1st of March 1901. It's just that their service has been constrained by the way they were able to live their lives, and to have that removed as a burden from our contemporary servicemen and women is a terrific thing. It makes us a better Army, Navy and Air Force. It makes us a better Defence Force; it makes us a better nation as a whole.[1]

Pride in Defence has traced the evolution of LGBTI service in Australia's military throughout the latter half of the twentieth century to the twenty-first century. From the introduction of the first explicit policy on homosexual military service during World War II, witch-hunts and persecution, through to the ADF's contemporary recognition that sexual and gender diversity makes for a stronger military, the trajectory has been remarkable. It might be unfinished, but the progress is still extraordinary.

The history of LGBTI service is one of determined courage and selflessness. Those who enlisted before they were legally permitted to do so performed their duty as required, making the many sacrifices that are necessary for life in the military—spending time away from friends and families, and forsaking the comforts taken for granted by most civilians. Most importantly, all made the sacred and firm commitment to defend Australia with their lives. Despite all of this, for most of the period covered in this book, simply being who they were was enough to see them discharged from the ADF. The legacy of being forced out of the military remains with many LGBTI veterans today.

LGBTI Defence members have had to contend with deeply ingrained institutionalised and embedded discrimination. Policies specifically targeting male homosexuality in the military emerged during World War II. Anxieties about women's sexuality in the post-war period saw officials in the WRANS, WRAAC and WRAAF embark on witch-hunts to investigate and expel women who were suspected of same-sex activity. By 1974, what had previously been an unwritten policy of persecution was formalised and entrenched in the ADF. Until the 1990s there was little official mention of transgender or intersex service, but oral histories reveal a long, silent history of transgender and intersex people serving in the ADF.

Despite obstacles, silences and persecution, LGB men and women in the services found each other. Indeed, a vibrant lesbian subculture thrived in the post-war women's services. The broader sweeping social change that took place in the 1970s also affected LGB people in the military. In the 1970s, gay and lesbian activists sporadically challenged the ban on LGB service. By the 1980s, LGB service members were daring to challenge the status quo. In 1992 politicians finally saw the merits of change and removed the LGB ban. In the aftermath of the ban being lifted, many LGB personnel remained closeted for fear of persecution. Others came out to varying responses and challenged the discriminatory policies that still existed. Real reform was achieved in December 2005 with the recognition of same-sex couples.

Just as there is a long and important history of LGB service, there is also a substantial history of transgender service—both before and after the removal of the transgender ban in September 2010. Intersex members have also made substantial contributions to the ADF, although the treatment of intersex personnel has differed in significant ways.

The contemporary ADF is making visible strides towards equality and inclusivity for LGBTI personnel. ADF members symbolically march in the Sydney Gay and Lesbian Mardi Gras in uniform. There is continuing work to be done, however, particularly to support transgender, non-binary and gender-diverse members.

The history of LGBTI service in the ADF shows the evolution of an institution regarded as one of Australia's most conservative. Unofficially, the

ADF's progression has depended on the people who have served—and continue to serve—in the organisation. By contributing, speaking out and being true to themselves, they have shown the deep significance of LGBTI service. The ADF's evolution has also been assisted by strong leadership who understand the value of diversity and inclusion.

The ADF's current approach towards diversity and inclusion is in line with policies adopted by culturally comparable armed forces. Karen D. Davis, who has explored the position of women in the contemporary Canadian military, states that the complexity of challenges that confront the modern military require 'increasing awareness of the importance of leveraging the maximum potential of all members, in particular those who contribute diversity to the overall capacity of the military'.[2] Similarly, David Rohall, Morten Ender and Michael Matthews argue that involving a wider group of American citizens in the armed forces and increasing social representation brings multiple perspectives and a larger array of skills and competencies to guide policies and strategic decisions.[3] The Ministry of Defence in the United Kingdom has also taken active steps to promote the benefits of diversity and inclusion in the armed forces. In 2007, for example, two publications were 'About defence: Defence and equality in the Armed Forces' and 'About defence: Homosexuality and the Armed Forces'.[4] Members of the British Armed Forces have been marching in London Pride in Uniform since 2008, and the Defence Diversity and Inclusion Strategy 2018–2030 includes recruiting more LGB people within its goals (although it is mostly silent about transgender people).[5] The inclusion of LGBTI personnel in international forces and the ADF not only enhances operational capability; it is also a reflection of broader shifts in civilian society towards the recognition of LGBTI rights, particularly those that have occurred in the new millennium.[6]

Governments in Canada and the United Kingdom have also acknowledged and apologised for repressive policies of the past that banned LGBTI people from serving. On 28 November 2017, Canadian Prime Minister Justin Trudeau delivered an emotional apology to all LGBTQ2 public servants and members of the military who were persecuted because of their sexuality or gender identity. He remarked:

> To those who were fired, to those who resigned, and to those who stayed at a great personal and professional cost; to those who wanted to serve, but never got the chance to because of who you are—you should have been permitted to serve your country, and you were stripped of that option. We are sorry. We were wrong. Indeed, all Canadians missed out on the important contributions you could have made to our society.

> You were not bad soldiers, sailors, airmen and women. You were not predators. And you were not criminals. You served your country with integrity, and veterans you are.
>
> You are professionals. You are patriots. And above all, you are innocent. And for all your suffering, you deserve justice, and you deserve peace. It is our collective shame that you were so mistreated. And it is our collective shame that this apology took so long—many who suffered are no longer alive to hear these words. And for that, we are truly sorry.[7]

The Canadian Government also awarded $110 million in compensation and personal letters of apology to more than 400 LGBTQ2 people persecuted under the so-called Gay Purge after a class action lawsuit.[8] Part of the settlement is also funding the construction of a monument in Ottawa to honour LGBTQ2 victims of the Gay Purge.[9]

In Britain, Defence Minister Johnny Mercer marked the twentieth anniversary of the United Kingdom lifting its ban on LGBT service by apologising to a group of LGBT veterans at a parliamentary reception in January 2020. He said, 'As the Minister for Defence, people and veterans, I wanted to personally apologise to you today for those experiences. Volunteering to serve is an act of bravery in itself; to volunteer for the chaotic, challenging nature of service life and yet within that community, which so many of us are proud of, experience discrimination of this sort is unacceptable.' Defence buildings across the United Kingdom were also lit up in rainbow colours to commemorate twenty years since the lifting of the ban.[10] To date, the Australian Government has refused calls for an apology to its own LGBTI ex-service members,[11] but the overseas precedents mean the push will only grow stronger.

Out of all the injustices LGBTI service personnel have faced, perhaps being written out of Australia's military history has been the most egregious. We hope this book, with a range of other emerging scholarship, will act as a corrective to this injustice and show just how important and extensive the LGBTI contributions have been to the ADF. In 2017, VCDF Ray Griggs provided a powerful Order of Australia ADF Oration. He asserted:

> I grew up, along with many of you tonight, in an ADF where it was illegal to be gay, where people were investigated, charged and dismissed from the Services (albeit that seemed to be arbitrary at times). I grew up in an ADF where a colleague committed suicide because he feared being 'outed'. It was a very different ADF to the one I see today, and frankly I know which ADF I want to be a part of—it's one where the focus is on what you bring to the

> fight, one that values the unique contribution you make, and one that does not exclude an individual because of their gender, race, religion or sexual orientation. Some observers of the ADF look at the aggregation of the social and cultural change and believe that we are lost in some sort of politically correct swamp, 'pandering to the progressives'. But the context here is not just about change in the ADF but the broader change in the society that we are for and of. Society continually changes and so must the ADF.[12]

In 2015 Squadron Leader Catherine Humphries of the RAAF also reflected on the ADF's journey:

> Over my eighteen years I've seen the military change a lot. I've seen things that are fundamental to the military still exist. So accepting and being more accepting of LGB hasn't stopped us being an effective force, hasn't caused issues on the front line. Now accepting transgender [people] hasn't caused any issues. It's not something that should be an issue. Hopefully.[13]

This book has shared the stories of individuals who were deeply courageous, not just because of their military service but also because they served, knowing that they were still not considered equal. Victimisation and prejudice have marred the experience of the ADF for far too many LGBTI people. However, the history of the ADF is one that also inspires hope. Today, LGBTI personnel can march proudly in uniform at Mardi Gras, the largest LGBTI event in Australia. This transformation must appear almost unimaginable to past generations of LGBTI service personnel. Yet it has shown that real change is possible and that when difference is accepted, institutions—and societies—thrive. LGBTI Defence members have always served. Now they can serve openly, with pride in who they are, knowing that they are part of many generations of LGBTI service personnel who have strengthened the ADF.

Notes

1 David Morrison, interview, 19 June 2017.

2 Karen D. Davis, 'Sex, gender and cultural intelligence in the Canadian Forces', *Commonwealth and Comparative Politics* 47, no. 4 (2009): 431.

3 David Rohall, Morten Ender and Michael Matthews, *Inclusion in the American Military: A Force for Diversity* (Lanham, MD: Lexington Books, 2017).

4 Ministry of Defence, 'About Defence: Diversity and equality in the armed forces', 2007; Ministry of Defence, 'About Defence: Homosexuality and the armed forces', 2007.

5 Ministry of Defence, 'A Force for Inclusion: Defence Diversity and Inclusion Strategy 2018–2030', https://assets.publishing.service.gov.uk/government/uploads/

system/uploads/attachment_data/file/746911/20180806-MOD_DI_Plan_A4_v14_Final-U.pdf, retrieved 10 June 2020; Tony Grew, 'Pride London's delight at men in uniform', *PinkNews*, 26 June 2008, https://www.pinknews.co.uk/2008/06/26/pride-londons-delight-at-men-in-uniform/, retrieved 10 June 2020.

6 See for example Reynolds and Robinson, *Gay and Lesbian, Then and Now*.

7 Justin Trudeau, 'Remarks by Prime Minister Justin Trudeau to apologize to LGBTQ2 Canadians', 28 November 2017, https://pm.gc.ca/en/news/speeches/2017/11/28/remarks-prime-minister-justin-trudeau-apologize-lgbtq2-canadians, retrieved 15 January 2020.

8 Olivia Chandler, '"A battle that we've won": LGBTQ military members get personal apologies', CBC, 25 November 2019, https://www.cbc.ca/news/canada/ottawa/lgbtq-military-canada-history-trudeau-apology-letter-1.5371444, retrieved 15 January 2020.

9 'Ottawa's newest national monument will honour LGBT Canadians and remember the "purge"', CBC, 25 January 2020, https://www.cbc.ca/news/politics/u-k-s-proposal-to-limit-huawei-s-role-in-5g-networks-most-pragmatic-path-expert-says-1.5437387/ottawa-s-newest-national-monument-will-honour-lgbt-canadians-and-remember-the-purge-1.5437395, retrieved 10 June 2020; Amanda Coletta, 'Canada to memorialize LGBT victims of Cold War-era "gay purge"', *Washington Post*, 3 May 2020, https://www.washingtonpost.com/world/the_americas/canada-trudeau-lgbt-gay-purge-memorial/2020/05/01/1223e302-888e-11ea-80df-d24b35a568ae_story.html, retrieved 10 June 2020.

10 Patrick Kelleher, 'British government finally apologises for banning gay people from armed forces, 20 years since ban was lifted', *PinkNews*, 10 January 2020, https://www.pinknews.co.uk/2020/01/10/gay-military-ban-armed-forces-apology-20-years-johnny-mercer-ministry-defence/, retrieved 15 January 2020; 'Ministry of Defence lit in rainbow colours to celebrate LGB personnel', 10 January 2020, https://www.gov.uk/government/news/mod-lit-in-rainbow-colours-to-celebrate-lgb-personnel--2, retrieved 15 January 2020.

11 Davey, '"I was absolutely shattered"'; Riseman, 'Why a national apology and redress for discharged LGBT service members matters'.

12 Ray Griggs, 'Order of Australia ADF Oration'.

13 Catherine Humphries, interview.

Bibliography

1. Archives and official sources

Australian Lesbian and Gay Archives, Melbourne

Bilney, Gordon, MP. 'Statement by Minister for Defence Science and Personnel.' 26 November 1992

'Confidential Australian Navy Order 35/66'. 6 July 1966

'Confidential Australian Navy Order 1-2/69'. 21 February 1969

Ex-RAAF policeman, letter to *OutRage*, 1991 (no date)

Queensland AIDS Council transgender archive. Notebook 1

Queensland AIDS Council transgender archive. Notebook 29

Australian Senate

Commonwealth Parliamentary Debates. Senate. 18 June 1992

Commonwealth Parliamentary Debates. Senate. 23 June 1992

Commonwealth Parliamentary Debates. Senate. 7 October 1992

Estimates Committee B. Report to the Senate. '1992–93 Senate Estimates Hearing Committee B—23 September 1992 Questions on Notice.'

Foreign Affairs, Defence and Trade Legislation Committee. Official Committee Hansard. 1 June 2015

—— Official Committee Hansard. 25 October 2017

Joint Standing Committee on Foreign Affairs, Defence and Trade (Defence Subcommittee). 'Inquiry into Military Justice Procedures in the Australian Defence Force.' Submission no. 46: Lucy Kardas, 6 May 1998

—— *Official Hansard Report*. 18 May 1998

—— 'Submissions to the Standing Committee on Foreign Affairs, Defence and Trade: Reference on Sexual Harassment in the Australian Defence Force.' 1993–94

Legal and Constitutional References Committee. 'Reference: Sexuality discrimination inquiry.' Adelaide, 16 August 1996

—— 'Reference: Sexuality discrimination inquiry.' Brisbane, 1 October 1996

—— 'Reference: Sexuality discrimination inquiry.' Canberra, 6 August 1996

—— Submissions to Sexuality Discrimination Inquiry, 1996

Senate Questions on Notice. 'Australian Defence Force: Same Sex Next of Kin.' Question 1678, 10 May 2000

Australian War Memorial

AWM15 18197

Department of Defence and Australian Defence Force publications

'ADFA Staff LGBTI Guide 2017'

Air Force Workforce Diversity. *Air Force Diversity Handbook: Transitioning Gender in Air Force.* April 2013

—— *Air Force Diversity Guide: For Lesbian, Gay, Bisexual Members*, September 2012

Department of Defence. 'Defence Diversity and Inclusion Strategy, 2012–2017.' https://www.defence.gov.au/Diversity/_Master/docs/DDIS-12-17.pdf. Retrieved 11 January 2020

—— Health Direction No. 210. 'Policy on the Prevention, Detection and Management of Blood Borne Virus Infection.' 13 June 2007

—— Health Directive 234. 'Medical Management of Gender dysphoria and Gender Realignment in Defence Members.' 13 April 2015

—— 'Understanding Transitioning Gender in the Workplace.' 10 November 2011

'Defence Census 2015 Public Report.' April 2016

Defence Health Manual Vol 02 Part 08. 2018

Defence Health Manual Vol 02 Part 09. 2019

Defence Instruction (General) PERS 15-3: 'Homosexual Behaviour in the Australian Defence Force.' 4 November 1985

Defence Instruction (General) PERS 16-16: 'Trans-gender Personnel in the Australian Defence Force.' 20 April 2000

Defence Instruction (General) PERS 21-1, AMDT no. 3: 'Political activities of Defence personnel.' 4 October 2007

Defence Instruction (General) PERS 50-1: 'Equity and Diversity in the Australian Defence Force.'

DEFGRAM No 553/2010. 'Trans-gender personnel in the Australian Defence Force.' 2 September 2010

DEFGRAM No 297/2016. 'Recognition of gender in Defence—Gender data collection and reporting for Australian Defence Force and Australian Public Service Employees.' 7 July 2016

Surgeon General Australian Defence Force. 'HIV Pre-Exposure Prophylaxis for ADF Members.' Health Bulletin No. 01/2017. 18 April 2017

National Archives of Australia

Allen, Brian. Service record. 311259. Central Army Records Office

Goldsby, Thomas. Service record. 311222. Central Army Records Office

NAA A161, 705/7/P1

NAA A451, 1974/6782

NAA A463, 1958/86

NAA A1209, 1977/503 PART 7

NAA A1209, 1992/3382

NAA A1813, 351/250/275

NAA A1813, 321/10/127

NAA A1813, 321/201/39

NAA A1813, 321/251/1

NAA A1814, 321/10/127

NAA A4090, 537/6/19
NAA A6122, 1194
NAA A6721, 1985/18156 PART 1
NAA A6721, 1985/18156 PART 2
NAA A6721, 1985/18156 PART 3
NAA A6721, 1985/18156 PART 4
NAA A6721, 1985/18156 PART 5
NAA A7481, 22
NAA A9177, 703/16/P1 PART 2
NAA A9737, 1992/13992 PART 1
NAA A9737, 1993/60684 PART 1
NAA A10756, LC1781 PART 2
NAA A11117, KCA535 PART 1
NAA A12909, 3954
NAA A12909, 3989
NAA A14217, 868
NAA A14217, 880
NAA B2458, 331048
NAA B4586, 704/1572/P1 PART 1
NAA B5679, 701/P1 PART 3
NAA BP899/6, 701/P1 PART 2
NAA BP1028/2, 701/P1 PART 2
NAA BP1028/2, 705/101/P1 PART 7
NAA BP1028/2, 705/101/P1 PART 10
NAA C475, 1718438
NAA C475, 1746037
NAA C1273, 705/P1
NAA C2439, 703/1/P1 PART 2
NAA K1259, A/1993/04651
NAA MP151/1, 445/201/123
NAA MP151/1, 455/201/234
NAA PP479/1, 705/1/P1 PART 4
NAA ST5533/1, 705/15/P1 PART 1
NAA ST5817/1, 705/24 P1 PART 1
NAA ST6999/1, 705/11/P1 PART 1
NAA ST7486/1, 705/RIC/P1 PART 1
Pollock, N. Service record. 0115708. Royal Australian Air Force

National Library of Australia

Robinson, Trevor. Interview with Clive Moore. Australian Lesbian and Gay Life Stories Oral History Project, 10 October 2013. NLA ORAL TRC 6510/4

State Library of New South Wales

Alexander 'Lex' Watson collection (courtesy Robert French)

Western Australia State Records Office

AU WA S1989—cons5527 1
AU WA S1989—cons5527 3

2. Unpublished sources

Oral history interviews

Allen, Brian. Interview with Noah Riseman. 19 November 2018. Melbourne
Amos, Scott. Interview with Noah Riseman. 28 August 2017. Townsville
Anderson, Cliff. Interview with Noah Riseman. 24 August 2017. Brisbane
'Anne'. Interview with Noah Riseman. 24 July 2015. Melbourne
'Archibald'. Interview with Noah Riseman. 9 December 2018. Gold Coast
Ashton, Sandy. Interview with Noah Riseman. 29 August 2015. Melbourne
Aulich, Terry. Interview with Noah Riseman. 14 May 2014, Melbourne
Bailey, Ed. Interview with Noah Riseman. 18 March 2017. Canberra
'Bill'. Interview with Noah Riseman. 7 June 2017. Melbourne
Billett, Graeme. Interview with Noah Riseman. 17 May 2018. Canberra
Blundell, Alix. Interview with Noah Riseman. 26 August 2017. Brisbane
Bone, Liam. Interview with Noah Riseman. 3 July 2017. Newcastle
Bowtell, Bill. Interview with Noah Riseman. 26 May 2017. Sydney
Bradford, David. Interview with Noah Riseman. 29 July 2015. Melbourne
Bradford, Riley. Interview with Noah Riseman. 20 June 2017. Canberra
'Brett'. Interview with Noah Riseman. 28 April 2017. Melbourne
Brown, Rob. Interview with Noah Riseman. 8 July 2017. Sydney
'Bruce'. Interview with Noah Riseman. 12 December 2018. Cairns
Bryant, Karl. Interview with Noah Riseman. 17 February 2016. Melbourne
Cahill, Craig. Interview with Noah Riseman. 28 September 2015. Launceston
Campbell, Max. Interview with Noah Riseman. 13 March 2015. Melbourne
Carlisle, Kate. Interview with Noah Riseman. 10 December 2017. Washington, DC
Carter, Janet. Interview with Shirleene Robinson. 9 August 2016. Perth
Chong, Vince. Interviews with Noah Riseman. 16 November 2016 and 28 November 2017. Canberra and Adelaide
Cincotta, Loretta. Interview with Noah Riseman. 1 August 2016. Canberra
Clark, Darran. Interview with Noah Riseman. 25 August 2017. Brisbane
Clinch, Bridget. Interview with Shirleene Robinson, 30 July 2015, Brisbane
Cone, Matt. Interview with Noah Riseman. 27 May 2017. Port Macquarie
Coogan, Eunice. Interview with Shirleene Robinson. 24 August 2016. Gold Coast
'Cooper'. Interview with Noah Riseman. 23 September 2016. Sydney
Cowin, Wally. Interview with Graham Willett. 13 September 2016. Brisbane
Croome, Rodney. Telephone interview with Noah Riseman. 14 December 2018
Crump, Elaine. Interview with Noah Riseman. 23 August 2017. Brisbane
D'Hagé, Adrian. Telephone interview with Noah Riseman. 24 April 2015
Daniels, Viv. Interview with Shirleene Robinson. 5 August 2016. Melbourne
'Diego'. Interview with Noah Riseman. August 2017
Dinnison, Michael. Interview with Noah Riseman. 30 August 2017. Townsville
Drew, Keith. Interview with Noah Riseman. 13 August 2016. Darwin
Duffield, Kyle. Interview with Noah Riseman. 19 April 2017. Bicheno, Tas
Duffy, Michael. Interview with Noah Riseman. 1 May 2014. Melbourne
Duniam, Shane. Interview with Noah Riseman. 29 May 2017. Lismore, NSW
Dunlop, John. Interview with Noah Riseman. 22 May 2017. Melbourne
Easterbrook, Leigh. Interview with Shirleene Robinson. 9 February 2017. Sydney
Edwards, Colin. Interview with Noah Riseman. 29 August 2017. Townsville
'Elliott'. Interview with Noah Riseman. 20 June 2017. Canberra

Fischer, Neal and Brychan Hawker. Interview with Graham Willett. 13 September 2016. Ipswich, Qld
Flanagan, Bob. Interview with Noah Riseman. 24 April 2018. Cessnock, NSW
Ford, Gen. Interview with Noah Riseman. 13 November 2014, London
Fry, Leon. Interview with Noah Riseman. 5 March 2016. Sydney
'Geoff'. Interview with Noah Riseman. 26 September 2016. Sydney
'Gina'. Interview with Noah Riseman. 30 May 2017. Sydney
Gleadhill, Garry. Interview with Noah Riseman. 20 April 2017. Branxholm, Tas
Gration, Richard. Interview with Noah Riseman. 13 December 2016. Sydney
Green, Becky. Interview with Noah Riseman. 21 September 2017. Sydney
Green, Lyndal. Interview with Shirleene Robinson. 1 July 2016. Ballarat
Griggs, Ray. Interview with Noah Riseman. 24 August 2018. Canberra
Haas, Connor. Interview with Noah Riseman. 20 June 2017. Canberra
Hamblin, Amy. Interviews with Noah Riseman. 26 August 2015 and 11 September 2015. Melbourne
Harding, Donna. Interview with Noah Riseman. 10 August 2015. Canberra
Heal, Catherine. Interview with Noah Riseman. 28 July 2017. Melbourne
'Helen'. Interview with Noah Riseman for transgender history project. 25 July 2019. Melbourne
Hendy, Julie. Interview with Shirleene Robinson. 13 May 2016. New South Wales
Humphries, Catherine. Interview with Noah Riseman. 11 August 2015. Canberra
'J'. Interview with Noah Riseman. 14 November 2016. Canberra
Jamieson, Graham. Interview with Noah Riseman. 7 July 2017, Berkeley Vale, NSW
'Jason'. Interview with Noah Riseman. 28 October 2016. Melbourne
Jefferies, Jennifer. Interview with Shirleene Robinson. 12 September 2016. Sydney
Jeffers, Keith. Interview with Noah Riseman. 16 November 2016. Canberra
'John'. Interview with Noah Riseman. 29 September 2015. Hobart
John M. Interview with Graham Willett. 18 January 2017. Wallan, Vic
Kardas, Lucy. Interview with Noah Riseman. 19 September 2017. Geelong
'Kate'. Interview with Noah Riseman. 31 July 2015. Melbourne
'Kevin'. Interview with Noah Riseman. 11 December 2017. Washington, DC
'Larissa'. Interview with Noah Riseman. 25 August 2017. Brisbane
Lawrie, Brian. Interview with Shirleene Robinson. 20 January 2016. Sydney
Lewis, Alan. Interview with Noah Riseman. 20 July 2015. Sunshine Coast
'Lexa'. Interview with Noah Riseman. 19 September 2016. Canberra
'Linda'. Interview with Noah Riseman. 31 July 2016. Canberra
Liversidge, Danny. Interview with Noah Riseman. 22 January 2018. Melbourne
Lord, Bill. Interview with Noah Riseman. 25 September 2017. Canberra
'Mark'. Interview with Noah Riseman. 4 March 2016. Sydney
Martin, Stuart. Interview with Graham Willett. 24 January 2017. Melbourne
McFarlane, Brian. Interview with Noah Riseman. 10 July 2015. Bowral, NSW
McInerney, Sandra. Interview with Shirleene Robinson. 8 December 2015. Canberra
McKeown, Kenton. Interview with Noah Riseman. 18 December 2018. Sydney
McLeod, Tony. Interview with Noah Riseman. 20 April 2017. Bicheno, Tas
'Mitch'. Interview with Noah Riseman. 9 August 2018. Adelaide
Mitchell, David. Interview with Noah Riseman. 10 July 2014. Brisbane
Montesini, Lorenzo. Interview with Noah Riseman. 2 March 2016. Sydney
Morrison, David. Interview with Noah Riseman. 18 November 2015. Melbourne
Morrison, Lieutenant General David. Interview with Noah Riseman. 19 June 2017. Canberra

Murray, Neil. Interview with Noah Riseman. 31 May 2017. Sydney
'Nathan'. Interview with Noah Riseman. 21 January 2016. Melbourne
'Nelson'. Interview with Noah Riseman. 19 December 2018
O'Brien, Stuart. Interview with Noah Riseman. 25 August 2018. Canberra
O'Sullivan, Kerry. Interview with Noah Riseman. 28 February 2018. Sydney
'Pencil'. Interview with Shirleene Robinson. 9 March 2017. Castlemaine, Vic
'Penelope'. Interview with Noah Riseman for transgender history project. 13 October 2018. Melbourne
'Pete'. Interview with Noah Riseman. 18 July 2018. Melbourne
'Peter'. Interview with Noah Riseman. 28 May 2017. Coffs Harbour, NSW
Pham, Dana. Interview with Noah Riseman. 24 May 2015. Sydney
Popham, Carole, and Christina Dennis. Interview with Shirleene Robinson. 18 March 2016. Melbourne
Pritchard, Adam. Interview with Noah Riseman. 2 September 2016. Melbourne
Quigley, Delia. Interview with Shirleene Robinson. 21 May 2017. Canberra
Ritchie, David. Interview with Noah Riseman. 7 July 2015. Sydney
Robertson, Jenny. Interview with Shirleene Robinson. 24 May 2017. Hobart
Robinson, Trevor. Interview with Noah Riseman. 18 July 2015. Brisbane
Sandra. Interview with Shirleene Robinson. 3 April 2017. Orbost, Vic
Seah, Mike. Interview with Noah Riseman. 13 November 2016. Canberra
Sillett, Yvonne. Interview with Shirleene Robinson. 29 August 2016. Daylesford, Vic
'Simon'. Interview with Noah Riseman. 9 September 2018. Melbourne
Sinclair, Chris. Interview with Noah Riseman. 10 September 2018. Melbourne
Spencer, Ronni. Interview with Noah Riseman. 3 August 2017. Melbourne
Strawbridge, Nola. Interviews with Shirleene Robinson. 15 October 2015 and 15 May 2016. Bribie Island, Qld, and Sydney
Struth, Susie. Interview with Shirleene Robinson. 18 February 2017. Melbourne
'Terri'. Interview with Noah Riseman. 19 October 2015. Sydney
Timmins, James. Interview with Shirleene Robinson. 17 August 2016. Sydney
Tolland, Russ. Interview with Noah Riseman. 24 August 2017. Brisbane
'Trent'. Interview with Noah Riseman. August 2017
van Netten, Anna. Interview with Shirleene Robinson. 6 April 2016, Newcastle, NSW
Wells, David. Interview with Noah Riseman. 12 August 2015. Canberra
Whelan, Tony. Interview with Noah Riseman. 2 May 2017. Melbourne
Williams, Richard. Interview with Noah Riseman. 29 November 2018. Griffith, NSW
Wilson, Joel. Interview with Noah Riseman. 15 May 2018. Canberra
'Xavier'. Interview with Noah Riseman. 10 August 2016. Melbourne

Personal papers

Blundell, Alix
Bryant, Karl
Carter, Janet
Clinch, Bridget
Cone, Matt
Duffield, Kyle
Gleadhill, Garry
Gration, Richard
Griggs, Ray
Kardas, Lucy
Mitchell, David

Seah, Mike
Sillett, Yvonne

Freedom of Information

Australian Human Rights Commission, Bridget Clinch Human Rights Commission case
Defence FOI 001/18/19
Defence FOI 009/17/18
Defence FOI 57/15/16
Defence FOI 212/15/16
Defence FOI 258/15/16
Defence FOI 358/17/18
Defence FOI 369/15/16

Legal cases

W.J. Carter QC. HREOC Unlawful Dismissal Case. 1995
C v. Commonwealth of Australia [2015] FCAFC 113
Chief of the Defence Force v Gaynor [2017] FCAFC 41
Commonwealth of Australia v Human Rights and Equal Opportunity Commission and Another [70 FCR 76] (1996)
Gaynor v Chief of the Defence Force (No. 3) [2015] FCA 1370; 237 FCR 188
X v. The Commonwealth of Australia and Another [200 CLR 177] (1999)

Melbourne eScholarship Research Centre

RMIT Social Policy Archive: Social Health Press Clippings Collection. Series 1—Social Health Press Clippings Collection. 'Transvestites/Transsexuals 5/76—12/79'
—— Series 1—Social Health Press Clippings Collection. 'Transvestites/Transsexuals 7/90—'

Other unpublished sources (including online)

Chong, Vince. 'Honouring all our diggers: Why we lay rainbow wreaths.' DEFGLIS. 2017. https://www.defglis.com.au/index.php/news/267-honouring-all-our-diggers-why-we-lay-rainbow-wreaths. Retrieved 13 January 2020
Commonwealth Superannuation Corporation. Fact Sheet MB03 07/13: Invalidity Benefits. https://www.csc.gov.au/members/advice-and-resources/factsheets-and-publications/militarysuper. Retrieved 15 June 2020
'Darlington Statement.' 10 March 2017. https://darlington.org.au/statement/. Retrieved 7 January 2020
'Defence Pride Network launched.' Australian Pride Network. 15 September 2015. http://australianpridenetwork.com.au/defence-pride-network-launched/. Retrieved 13 January 2020
'DEFGLIS relaunches for 2012.' DEFGLIS. http://pandora.nla.gov.au/pan/132273/20120321-1646/www.defglis.com/index.php/en/media-releases/68-launch-of-new-website-2.html. Retrieved 11 January 2020
Downer, Alexander. 'Media release: The Defence Force and homosexuals.' 23 November 1992
Fischer, Tim. 'Media release: Homosexuals in the Defence Force.' 23 November 1992
'GLAAD media reference guide—Transgender.' GLAAD. https://www.glaad.org/reference/transgender. Retrieved 15 May 2019

'Intersex population figures.' Intersex Human Rights Australia. https://ihra.org.au/16601/intersex-numbers/. Retrieved 7 January 2020

Keating, Paul. 'Statement by the Prime Minister, the Hon P.J. Keating, MP, Australian Defence Force policy on homosexuality.' 23 November 1992

'Ministry of Defence lit in rainbow colours to celebrate LGB personnel.' 10 January 2020. https://www.gov.uk/government/news/mod-lit-in-rainbow-colours-to-celebrate-lgb-personnel--2. Retrieved 15 January 2020

Ministry of Defence (UK). 'A Force for Inclusion: Defence Diversity and Inclusion Strategy 2018–2030'. https://assets.publishing.service.gov.uk/government/uploads/system/uploads/attachment_data/file/746911/20180806-MOD_DI_Plan_A4_v14_Final-U.pdf. Retrieved 10 June 2020

New South Wales. 'Inquiry into gay and transgender hate crimes between 1970 and 2010.' https://www.parliament.nsw.gov.au/committees/inquiries/Pages/inquiry-details.aspx?pk=2510#tab-members. Retrieved 24 February 2019

'Service warrant officers to march at Mardi Gras 2015.' 11 February 2015. https://www.defglis.com.au/index.php/news/168-service-warrant-officers-to-march-at-mardi-gras-2015. Retrieved 2 July 2015

State Coroner's Court of New South Wales. 'Inquest into the death of Scott Russell Johnson.' 30 November 2017. http://www.coroners.justice.nsw.gov.au/Documents/Scott%20Johnson%20findings%20%20final.pdf. Retrieved 24 February 2019

Trudeau, Justin. 'Remarks by Prime Minister Justin Trudeau to apologize to LGBTQ2 Canadians.' 28 November 2017. https://pm.gc.ca/en/news/speeches/2017/11/28/remarks-prime-minister-justin-trudeau-apologize-lgbtq2-canadians. Retrieved 15 January 2020

'What is intersex?' Intersex Human Rights Australia. https://ihra.org.au/18106/what-is-intersex/. Retrieved 2 January 2020

World Professional Association for Transgender Health. 'Position statement on medical necessity of treatment, sex reassignment, and insurance coverage in the USA.' 21 December 2016. https://www.wpath.org/newsroom/medical-necessity-statement. Retrieved 9 January 2020

3. Books, chapters, articles and theses

Abrams, Lynn. *Oral History Theory.* London and New York: Routledge, 2010

Agostino, Katerina. 'Men, identity and military culture.' In *Male Trouble: Looking at Australian Masculinities*, ed. Stephen Tomsen and Mike Donaldson, pp. 108–31. North Melbourne, Vic: Pluto Press, 2003

Altman, Dennis. 'It began by accident.' In *Mardi Gras! True Stories*, ed. Richard Wherrett, pp. 8–10. Ringwood, Vic: Viking, 1999

Anderson, David. *The Challenge of Military Service: Defence Personnel's Conditions in a Changing Social Context.* Background Paper 5, 1997–98. Canberra: Australian Parliamentary Library, 1997

Anderson, Roderic. *Free Radical: A Memoir of a Gay Political Activist.* Salisbury, Qld: Lulu, 2006

Arrow, Michelle. *The Seventies: The Personal, the Political, and the Making of Modern Australia.* Sydney: NewSouth, 2019

—— '"These are just a few examples of our daily oppressions": Speaking and listening to homosexuality in Australia's Royal Commission on Human Relationships, 1974–1977.' *Journal of the History of Sexuality* 27, no. 2 (2018): 234–63

Australia. 'Report of Case Study No. 40: The response of the Australian Defence Force to allegations of child sexual abuse.' Sydney: Royal Commission into Institutional Responses to Child Sexual Abuse, August 2017. https://www.childabuseroyalcommission.gov.au/sites/default/files/case_study_40_-_findings_report_-_australian_defence_force.pdf

Australia. Senate Committee on Foreign Affairs, Defence and Trade. 'Sexual Harassment in the Australian Defence Force.' August 1994

Australian Human Rights and Equal Opportunity Commission. *Report of the Human Rights and Equal Opportunity Commission on Australian Defence Force Policy on Homosexuality.* Sydney: Craft Printing Industries Pty Ltd, 1992

Beck, Kristin, and Anne Speckhard. *Warrior Princess: A US Navy Seal's Journey to Coming Out Transgender.* McLean, VA: Advances Press, 2013

Belkin, Aaron, and Jason McNichol. 'The effect of including gay and lesbian soldiers in the Australian Defence Forces: Appraising the evidence.' Center for the Studies of Sexual Minorities in the Military, University of California, 19 September 2000. http://www.dont.law.stanford.edu/wp-content/uploads/2010/11/Australian_Defence_Forces.pdf

Benjamin, Harry. 'Transsexualism and transvestism as psycho-somatic and somato-psychic syndromes.' In *The Transgender Studies Reader*, ed. Susan Stryker and Stephen Whittle, pp. 45–52. New York: Routledge, [1954] 2006

Berg, Charles, MD, and Clifford Allen, MD. *The Problem of Homosexuality.* New York: Citadel Press, 1958

Berman, Alan, and Shirleene Robinson. *Speaking Out: Stopping Homophobic and Transphobic Abuse in Queensland.* Bowen Hills, Qld: Australian Academic Press, 2010

Bérubé, Allan. *Coming Out Under Fire: The History of Gay Men and Women in World War Two.* New York: Free Press, 1990

Blair, Anne. *Ruxton: A Biography.* Crows Nest, NSW: Allen & Unwin, 2004

Blaxland, John (ed.). *East Timor Intervention: A Retrospective on INTERFET.* Carlton, Vic: Melbourne University Publishing, 2015

Blewett, Neal. *A Cabinet Diary: A Personal Record of the First Keating Government.* Kent Town, SA: Wakefield Press, 1999

Bomford, Janette. *Soldiers of the Queen: Women in the Australian Army.* South Melbourne: Oxford University Press, 2001

Bourne, Stephen. *Fighting Proud: The Untold Story of the Gay Men Who Served in Two World Wars.* London: I.B. Taurus, 2017

Boyd, Nan Alamilla. 'Who is the subject? Queer theory meets oral history.' *Journal of the History of Sexuality* 17, no. 2 (May 2008): 177–89

Bradford, David. *The Gunners' Doctor: Vietnam Letters.* North Sydney: Random House Australia, 2007

—— *Tell Me I'm Okay: A Doctor's Story.* Clayton, Vic: Monash University Publishing, 2018

Brown, Gary. 'Homosexuality and the Australian Defence Force: The issues.' Background Paper Number 16, 1992. Canberra: Parliamentary Research Service, 14 August 1992

Brown, George R. 'Transsexuals in the military: Flight into hypermasculinity.' [In English]. *Archives of Sexual Behavior* 17, no. 6 (1988): 527–37

Brown, George R., and Kenneth T. Jones. 'Mental health and medical health disparities in 5135 transgender veterans receiving healthcare in the Veterans Health Administration: A case-control study.' *LGBT Health* 3, no. 2 (2016): 122–31

Brown, James. *Anzac's Long Shadow: The Cost of Our National Obsession.* Collingwood, Vic: Redback, 2014

Brown, Mildred L., and Chloe Ann Rounsley. *True Selves: Understanding Transsexualism—for Families, Friends, Coworkers, and Helping Professionals.* San Francisco: Jossey-Bass, 1996

Buhrich, Neil. 'A heterosexual transvestite club: Psychiatric aspects.' *Australian and New Zealand Journal of Psychiatry* 10, no. 4 (1976): 331–5

Bull, Donna. 'Mobility: What effect does it have on ADF personnel?' *Australian Defence Force Journal* 138 (September–October 1999): 31–7

CAMP NSW. *Homosexuals and Human Relationships: Submission to the Royal Commission on Human Relationships.* Sydney: CAMP NSW, September 1975

Capezza, Nicole. 'Homophobia and sexism: The pros and cons to an integrative approach.' *Integrative Psychological and Behavioral Science* 41, no. 3 (2007): 248–53

Clausewitz, Carl von. *On War*. Trans. Michael Howard and Peter Paret. Princeton, NJ: Princeton University Press, 1976

Collyer, David C. *From Fishnets to Foxholes and Back Again*. Ballarat, Vic: David C. Collyer, 2001

Commonwealth of Australia. *Royal Commission on Human Relationships Final Report*. Vol. 1, Canberra: Australian Government Publishing Service, 1977

—— *Royal Commission on Human Relationships Final Report*. Vol. 5, Canberra: Australian Government Publishing Service, 1977

Commonwealth Ombudsman. *Commonwealth Ombudsman and Defence Force Ombudsman Annual Reports 1988–89*. Canberra: Australian Government Publishing Service, 1989

—— *Commonwealth Ombudsman Fourth Annual Report 1980–81*. Canberra: Australian Government Publishing Service, 1981

—— *Defence Force Ombudsman: Twenty-five Years of Service*. Canberra: Commonwealth Ombudsman, 2009. https://www.ombudsman.gov.au/__data/assets/pdf_file/0018/36207/defence_force_ombudsman-twenty-five_years_of_service.pdf

Cooper, Alastair. '1955–1972: The era of Forward Defence.' In *The Royal Australian Navy: A History*, ed. David Stephens, pp. 181–209. South Melbourne: Oxford University Press, 2001

Cornelisse, Vincent J., Christopher K. Fairley, Mark Stoove et al. 'Evaluation of Preexposure (PrEP) eligibility criteria, using sexually transmissible infections as markers of human immunodeficiency (HIV) risk at enrollment in PrEPX, a large Australian HIV PrEP trial.' *Clinical Infectious Disease* 67, no. 12 (2018): 1847–52

Crotty, Martin, and Mark Edele. 'Total war and entitlement: Towards a global history of veteran privilege.' *Australian Journal of Politics and History* 59, no. 1 (March 2013): 15–32

Cummings, Katherine. *Katherine's Diary: The Story of a Transsexual*. Rev. edn. Tascott, NSW: Beaujon Press, 2007

Curtis-Otter, Margaret. *WRANS: The Women's Royal Australian Naval Service*. Sydney: Naval Historical Society of Australia, 1996

Dapin, Mark. *The Nashos' War: Australian National Servicemen and Vietnam*. Melbourne: Viking, 2014

—— *R&R*. Melbourne: Penguin Books, 2015

Darian-Smith, Kate. 'War and Australian society.' In *Australia's War, 1939–45*, ed. Joan Beaumont, pp. 45–68. Sydney: Allen & Unwin, 1996

Davis, Karen D. 'Sex, gender and cultural intelligence in the Canadian forces.' *Commonwealth and Comparative Politics* 47, no. 4 (2009): 430–55

Defence Abuse Response Taskforce. *Final Report*. Canberra: Commonwealth of Australia, March 2016

—— *Report on Abuse in Defence*. Canberra, 2014

Deslandes, Paul R. 'Situational Homosexuality.' In *Global Encyclopedia of Lesbian, Gay, Bisexual, Transgender, and Queer (LGBTQ) History*, ed. Howard Chiang, Anjali Arondekar, Marc Epprecht, Jennifer Evans, Ross G. Forman, Hanadi Al-Samman, Emily Skidmore and Zeb Tortorici, pp. 1501–7. Farmington Hills, MI: Charles Scribner's Sons, 2019

Eastman, Edward, Robert P. Archer and J.D. Ball. 'Psychosocial and life stress characteristics of Navy families: Family Environment Scale and Life Experiences Scale finding.' *Military Psychology* 2, no. 2 (1990): 113–27

Eisinger, Robert W., Carl W. Diettenbach and Anthony S. Fauci. 'HIV viral load and transmissibility of HIV infection.' *Journal of the American Medical Association* 321, no. 5 (2019): 451–2

Ekins, Ashley. 'Vietnam: A winnable war?' In *War Strategy and History: Essays in Honour of Professor Robert O'Neill*, ed. Daniel Marston and Tamara Leahy, pp. 5–30. Canberra: ANU Press, 2016

Enloe, Cynthia. *Maneuvers: The International Politics of Militarizing Women's Lives*. Berkeley, CA: University of California Press, 2000

Estes, Steve. *Ask and Tell: Gay and Lesbian Veterans Speak Out*. Chapel Hill, NC: University of North Carolina Press, 2007

—— 'Ask and tell: Gay veterans, identity and oral history on a civil rights frontier.' *Oral History Review* 32, no. 2 (2005): 21–47

Evans, Raymond, and Kay Saunders. 'The evolution of the Australian housewife.' In *Gender Relations in Australia: Domination and Negotiation*, ed. Kay Saunders and Raymond Evans, pp. 175–96. Sydney: Harcourt, Brace & Company, 1992

Ford, Ruth. 'Disciplined, punished and resisting bodies: Lesbian women and the Australian armed services, 1950s/60s.' *Lilith*, no. 9 (1996): 53–77

—— 'Lesbians and loose women: Female sexuality and the women's services during World War II.' In *Gender and War: Australians at War in the Twentieth Century*, ed. Joy Damousi and Marilyn Lake, pp. 81–104. Cambridge: Cambridge University Press, 1995

Frame, Tom. *Where Fate Calls: The HMAS Voyager Tragedy*. Sydney: Hodder & Stoughton, 1992

Frank, Nathaniel. *Unfriendly Fire: How the Gay Ban Undermines the Military and Weakens America*. New York: Thomas Dunne Books, 2009

French, Robert. '"Persons with serious character defects": Homosexuals in the Commonwealth Public Service, 1953–1974.' In *Intimacy, Violence and Activism: Gay and Lesbian Perspectives on Australasian History and Society*, ed. Graham Willett and Yorick Smaal, pp. 119–32. Clayton, Vic: Monash University Publishing, 2013

Gibson, E. Lawrence. *Get Off My Ship: Ensign Berg vs. The US Navy*. New York: Avon Books, 1978

Goldstein, Joshua. *War and Gender: How Gender Shapes the War System and Vice Versa*. Cambridge: Cambridge University Press, 2001

Green, Anna. 'Individual remembering and "collective memory": Theoretical presuppositions and contemporary debates.' *Oral History* 32, no. 2 (2004): 35–44

Grogan, Peter, and Rodney Croome. 'A submission to the ALP Caucus Committee on the Australian Defence Force policy on homosexuality.' 1992

Hadrian, 'Homosexuality in the forces', *Campaign Circular* (November–December 1972), reprinted in *CAMP Ink*, issue 3 (1973)

Ham, Paul. *Vietnam: The Australian War*. Pymble, NSW: HarperCollins, 2007

Hardy, David. *Bold: Stories from Older Lesbian, Gay, Bisexual, Transgender and Intersex People*. Panton Hill, Vic: Rag & Bone Man Press, 2015

Harris, Stanley E. 'Military policies regarding homosexual behavior.' *Journal of Homosexuality* 21, no. 4 (1991): 67–74

Hart, Gavin. 'Sexual behaviour in a war environment.' *Journal of Sex Research* 11, no. 3 (1975): 218–26

Haskins, Victoria. 'The girl who wanted to go to war: Female patriotism and gender construction in Australia's Great War.' *History Australia* 14, no. 2 (2017): 169–86

Henry, P.J., and Curtis D. Hardin. 'The contact hypothesis revisited.' *Psychological Science* 17, no. 10 (2006): 862–8

Hippler, Mike. *Matlovich, the Good Soldier.* Boston: Alyson Publications, 1989

Hirschfeld, Magnus. *Transvestites: The Erotic Drive to Cross Dress.* Trans. Michael A. Lombardi-Nash. New York: Prometheus Books, [1910] 1991

Hodge, Dino. *Did You Meet Any Malagas? A Homosexual History of Australia's Tropical Capital.* Nightcliff, NT: Little Gem Publications, 1993

Holbrook, Carolyn. *Anzac: The Unauthorised Biography.* Sydney: NewSouth, 2014

Humphrey, Mary Ann. *My Country, My Right to Serve: Experiences of Gay Men and Women in the Military, World War II to the Present.* New York: Harper Collins, 1990

Jackson, Paul. *One of the Boys: Homosexuality in the Military During World War II.* Montreal: McGill University Press, 2004

Janowitz, Morris. 'Military institutions and citizenship in Western societies.' *Armed Forces and Society* 2, no. 2 (1976): 185–204

Jennings, Rebecca. '"It was quite a scary time": Lesbians and violence in post-war Australia.' In *Gender Violence in Australia: Historical Perspectives*, ed. Alana Piper and Ana Stevenson, pp. 145–59. Clayton, Vic: Monash University Publishing, 2019

—— *Unnamed Desires: A Sydney Lesbian History.* Clayton, Vic: Monash University Publishing, 2015

Jericho, Jyonah. 'Hegemonic masculinity in the Australian Defence Force—the exclusion of women from combat service as state policy, 1973–2013.' PhD thesis, University of Sydney, 2015

Jones, Craig (ed.). *Fighting with Pride: LGBTQ in the Armed Forces.* Barnsley, UK: Pen and Sword Military, 2019

Jones, Peter. 'Towards self-reliance.' In *The Royal Australian Navy: A History*, ed. David Stevens, pp. 211–37. South Melbourne: Oxford University Press, 2001

Junger, Sebastian. *Tribe: On Homecoming and Belonging.* London: HarperCollins Publishers, 2016

Kahn, James O., Jeffrey N. Martin, Michelle E. Roland et al. 'Feasibility of Postexposure Prophylaxis (PEP) against human immunodeficiency viral infection after sexual or injection drug use exposure: The San Francisco PEP study.' *Journal of Infectious Diseases* 183 (2001): 707–14

Kansteiner, Wolf. 'Finding meaning in memory: A methodological critique of collective memory studies.' *History and Theory* 41, no. 2 (2002): 179–97

Kerry, Stephen. 'Australian news media's representation of Cate McGregor, the highest ranking Australian transgender military officer.' *Journal of Gender Studies* 27, no. 6 (2018): 683–93

Krebs, Ronald R. *Fighting for Rights: Military Service and the Politics of Citizenship.* Ithaca, NY: Cornell University Press, 2006

Lake, Marilyn, and Henry Reynolds. *What's Wrong with Anzac? The Militarisation of Australian History.* Sydney: UNSW Press, 2010

Lapovsky Kennedy, Elizabeth. 'Telling tales: Oral history and the construction of pre-Stonewall lesbian history.' In *The Oral History Reader*, ed. Robert Perks and Alistair Thomson, pp. 344–55. London: Routledge, 1998

Lapovsky Kennedy, Elizabeth, and Madeline D. Davis. *Boots of Leather, Slippers of Gold: The History of a Lesbian Community.* New York and London: Routledge, 1993

Lopez, Dominic. 'Brothers and sisters in arms: Experiences of gay soldiers in the Australian Army.' *Australian Army Journal* 10, no. 3 (2013): 213–25

Lord, James. *My Queer War.* New York: Farrar, Straus & Giroux, 2010

Lothstein, Leslie Martin. *Female-to-Male Transsexualism: Historical, Clinical, and Theoretical Issues.* Boston: Routledge & Kegan Paul, 1983

Lummis, Trevor. 'Structure and validity in oral evidence.' In *The Oral History Reader*, ed. Robert Perks and Alistair Thomson. London and New York: Routledge, 2006

Lupton, Deborah. *Moral Threats and Dangerous Desires: AIDS in the News Media*. London: Taylor & Francis, 1994

Marcus, Eric (ed.). *Making History: The Struggle for Gay and Lesbian Equal Rights, 1945–1990, an Oral History*. New York: HarperCollins, 1992

Marsh, Ian, and Larry Galbraith. 'The political impact of the Sydney Gay and Lesbian Mardi Gras.' *Australian Journal of Political Science* 30, no. 2 (1995): 300–20

Maynard, John. 'The rise of the modern Aboriginal political movement, 1924–39.' In *Serving Our Country: Indigenous Australians, War, Defence and Citizenship*, ed. Joan Beaumont and Allison Cadzow, pp. 113–34. Sydney: NewSouth, 2018

McFarlane, Brian. *We Band of Brothers: A True Australian Adventure Story*. Newcastle, NSW: Exact Print + Design, 2000

McKenna, Mark. 'Anzac Day: How did it become Australia's national day?' In *What's Wrong with Anzac? The Militarisation of Australian History*, ed. Marilyn Lake and Henry Reynolds, pp. 110–34. Sydney: NewSouth, 2010

Merritt, Rich. *Secrets of a Gay Marine Porn Star*. New York: Kensington Books, 2005

Mezey, Susan Gluck. *Transgender Rights: From Obama to Trump*. New York: Routledge, 2019

Ministry of Defence (UK). 'About Defence: Diversity and equality in the armed forces.' 2007

—— 'About Defence: Homosexuality and the armed forces.' 2007

Montesini, Lorenzo. *My Life and Other Misdemeanours*. Ringwood, Vic: Viking, 1999

Morris, Ben, and Noah Riseman. 'Volunteers with a legal impediment: Australian National Service and the question of overseas service in Vietnam.' *History Australia* 16, no. 2 (2019): 266–86

Moskos Jr, Charles. 'From citizen's army to social laboratory.' *Wilson Quarterly* 17 (winter 1993): 83–94

O'Brien, Kerry. *Keating*. Sydney: Allen & Unwin, 2015

Paige, Caroline. *True Colours: My Life as the First Openly Transgender Officer in the British Armed Forces*. London: Biteback Publishing, 2017

Parco, James, and David Levy (eds). *Evolution of Government Policy Towards Homosexuality in the US Military: The Rise and Fall of DADT*. Abingdon, UK, and New York: Routledge, 2014

Phelps, Lieutenant Colonel M.L. 'The Australian Army's culture: From institutional warrior to pragmatic professional.' *Australian Defence Force Journal* 123 (March/April 1997): 37–43

Plummer, Ken. *Telling Sexual Stories: Power, Change and Social Worlds*. London and New York: Routledge, 1995

Polchar, Joshua, Tim Sweijs, Philipp Marten, and Jan Galdiga. *LGBT Military Personnel: A Strategic Vision for Inclusion*. The Hague: The Hague Centre for Strategic Studies, 2014

Popular Memory Group. 'Popular memory: Theory, politics, methods.' In *The Oral History Reader*, ed. Robert Perks and Alistair Thomson, pp. 43–53. London and New York: Routledge, 2006

Prendergast, Gavin. 'Unacceptable sexual behaviour—the real victims.' *Australian Defence Force Journal*, no. 102 (September–October 1993): n.p.

Quinn, Major Kathryn. 'Sexual harassment in the Australian Defence Force.' Canberra: Director Publishing and Visual Communications, Defence Centre, December 1996

Rawson, Kelly Jacob, and Cristan Williams. 'Transgender*: The rhetorical landscape of a term.' *Present Tense* 3, no. 2 (2014): 1–9

Reghenzani, Christine. 'Women in the ADF: Six decades of policy change (1950 to 2011).' Australian Parliamentary Library Summer Scholar's Paper. https://www.aph.gov.au/~/

media/05%20About%20Parliament/54%20Parliamentary%20Depts/544%20Parliamentary%20Library/pubs/apf/scholarship/CReghenzani.pdf?la=en

Report of the Caucus Joint Working Group on Homosexual Policy in the Australian Defence Force (ADF). September 1992

Reynolds, Robert. *From Camp to Queer: Re-making the Australian Homosexual*. Melbourne: Melbourne University Press, 2002

—— *What Happened to Gay Life?* Sydney: NewSouth, 2007

Reynolds, Robert, and Shirleene Robinson. *Gay and Lesbian, Then and Now: Australian Stories from a Social Revolution*. Carlton, Vic: Black Inc., 2016

Riseman, Noah. 'Activism and Australia's ban on gay, lesbian and bisexual military service in the 1970s–80s.' *Australian Feminist Studies* 33, no. 95 (2018): 147–63

—— 'Ex-service activism after 1945.' In *Serving Our Country: Indigenous Australians, War, Defence and Citizenship*, ed. Joan Beaumont and Allison Cadzow, pp. 261–81. Sydney: NewSouth, 2018

—— 'From witch-hunts to pride balls: The ADF and LGBTI service personnel.' In *Beyond Combat: Australian Military Activity Away from the Battlefield*, ed. Tristan Moss and Tom Richardson, pp. 29–44. Sydney: NewSouth, 2018

—— 'Hunting gays and lesbians in the Australian Defence Force, 1974–1992.' *Journal of the History of Sexuality* 28, no. 3 (2019): 325–56

—— 'The interview–document nexus: Recovering histories of LGBTI military service in Australia.' *Archivaria* 87 (Spring 2019): 6–33

—— '"Just another start to the denigration of Anzac Day": Evolving commemorations of LGBTI military service.' *Australian Historical Studies* 48, no. 1 (2017): 35–51

—— 'Outmanoeuvring defence: The Australian debates over gay and lesbian military service, 1992.' *Australian Journal of Politics and History* 61, no. 4 (2015): 562–75

—— 'Transgender inclusion and Australia's failed Sexuality Discrimination Bill.' *Australian Journal of Politics and History* 65, no. 2 (2019): 259–77

—— 'Transgender policy in the Australian Defence Force: Medicalization and its discontents.' *International Journal of Transgenderism* 17, no. 3–4 (2016): 141–54

Riseman, Noah, and Richard Trembath. *Defending Country: Aboriginal and Torres Strait Islander Military Service since 1945*. St Lucia, Qld: University of Queensland Press, 2016

Riseman, Noah, Shirleene Robinson, and Graham Willett. *Serving in Silence? Australian LGBT Servicemen and Women*. Sydney: NewSouth, 2018

Robinson, Shirleene. 'A military discharge certificate.' In *Queer Objects*, ed. Chris Brickell and Judith Collard, pp. 297–301. Dunedin: Otago University Press, 2019

—— 'Sexuality at a cost: Lesbian servicewomen in the Australian military, 1960s–1980s.' In *Beyond Combat: Australian Military Activity Away from the Battlefield*, ed. Tristan Moss and Tom Richardson, pp. 45–58. Sydney: NewSouth, 2018

Rohall, David, Morten Ender and Michael Matthews. *Inclusion in the American Military: A Force for Diversity*. Lanham, MD: Lexington Books, 2017

Roque Ramirez, Horacio N., and Nan Alamilla Boyd. 'Introduction: Close encounters. The body and knowledge in queer oral history.' In *Bodies of Evidence: The Practice of Queer Oral History*, ed. Nan Alamilla Boyd and Horacio N. Roque Ramirez, pp. 1–20. Oxford: Oxford University Press, 2012

Rumble, Gary, Melanie McKean, and Dennis Pearce. *Report of the Review of Allegations of Sexual and Other Abuse in Defence: Facing the Problems of the Past*. Vol. 1, Canberra: Commonwealth of Australia, 2011

Salter, J.S. *2/95 Soldier Attitude and Opinion Survey: Equal Employment Opportunity Issues*. Canberra: 1st Psychological Research Unit, Australian Army, May 1996

—— *1995 Officer Attitude and Opinion Survey: Equal Employment Opportunity Issues*. Canberra: 1st Psychological Research Unit, Australian Army, February 1996

—— *1995 Soldier Attitude and Opinion Survey: Equal Employment Opportunity Issues.* Canberra: 1st Psychological Research Unit, Australian Army, February 1996

Salter, J.S., and S.E. Hodson. *Soldier Attitude and Opinion Survey: Topical Items—1992.* Canberra: 1st Psychological Research Unit, Australian Army, September 1995

Santos, Ana Lúcia. 'Beyond binarism? Intersex as an epistemological and political challenge.' *RCCS Annual Review*, no. 6 (2014): 123–40

Scates, Bruce. *A Place to Remember: A History of the Shrine of Remembrance.* Cambridge: Cambridge University Press, 2009

Seal, Graham. *Inventing Anzac: The Digger and National Mythology*. St Lucia, Qld: University of Queensland Press in association with the API Network and Curtin University of Technology, 2004

Segal, David R., Paul A. Gade and Edgar M. Johnson. 'Homosexuals in Western armed forces.' *Society* 31, no. 1 (1993): 37–42

Shilts, Randy. *Conduct Unbecoming: Lesbians and Gays in the US Military, Vietnam to the Persian Gulf.* London: Penguin Books, 1993

Smaal, Yorick. *Sex, Soldiers and the South Pacific, 1939–45: Queer Identities in Australia in the Second World War.* London: Palgrave Macmillan, 2015

Smaal, Yorick, and Graham Willett. 'Eliminate the "females": The New Guinea affair and medical approaches to homosexuality in the Australian Army in the Second World War.' In *The Pacific War: Aftermaths, Remembrance and Culture*, ed. Christina Twomey and Ernest Koh, pp. 233–50. Abingdon, UK, and New York: Routledge, 2015

Smith, Hugh. 'The dynamics of social change and the Australian Defence Force.' *Armed Forces and Society* 21, no. 4 (1995): 531–51

—— *Homosexuality and the Australian Defence Force: Individual Rights vs Organizational Realities*. Australian Defence Studies Centre Working Paper. Campbell, ACT: University College, University of New South Wales, Australian Defence Force Academy, 1992

Snyder-Hill, Stephen. *Soldier of Change: From the Closet to the Forefront of the Gay Rights Movement*. Lincoln, NE: University of Nebraska Press, 2014

Spurling, Kathryn. '1991–2001: The era of defence reform.' In *The Royal Australian Navy*, ed. David M. Stevens, pp. 269–94. South Melbourne: Oxford University Press, 2001

—— 'From exclusion to submarines: The integration of Australian naval women volunteers.' *Australian Defence Force Journal*, no. 139 (November/December 1999): 34–40

Stanley, Peter. *Bad Characters: Sex, Crime, Mutiny, Murder and the Australian Imperial Force.* Millers Point, NSW: Pier 9, 2010

Stewart, Ngaere B. '"Married" to the military: The experiences of partners of Australian Defence Force members.' BA Honours Thesis, Edith Cowan University, 2004

Stirling, Peter. *So Different: An Extraordinary Autobiography*. Sydney: Simon & Schuster Australia, 1989

Stryker, Susan. *Transgender History: The Roots of Today's Revolution*. 2nd edn. New York: Seal Press, 2017

Summerfield, Penny. 'Culture and composure: Creating narratives of the gendered self in oral history interviews.' *Cultural and Social History* 1, no. 1 (2004): 65–93

Tallberg, Teemu, and Johanna Valenius. 'Men, militaries and civilian societies in interaction.' *Norma: Nordic Journal for Masculinity Studies* 3, no. 2 (2008): 85–98

Thomson, Alistair. *Anzac Memories: Living with the Legend*. 2nd edn. Clayton, Vic: Monash University Publishing, 2013

—— 'Anzac memories: Putting popular memory theory into practice in Australia.' In *The Oral History Reader*, ed. Robert Perks and Alistair Thomson, pp. 244–54. London and New York: Routledge, 2006

—— 'Anzac memories revisited: Trauma, memory and oral history.' *Oral History Review* 42, no. 1 (2015): 1–29

Tovey, Noel. *Little Black Bastard: A Story of Survival*. Sydney: Hodder Headline, 2005
Transgender Studies Quarterly 5, no. 4 (2018)
Twomey, Christina. 'Trauma and the reinvigoration of Anzac: An argument.' *History Australia* 10, no. 3 (2013): 85–108
'United Nations Human Rights Committee (UNHRC): Views on Edward Young v. Australia.' *International Legal Materials* 42, no. 5 (2003): 1244–57
Ustinoff, Julie. 'Homebodies and weekend handymen.' In *The 1960s in Australia: People, Power and Politics*, ed. Shirleene Robinson and Julie Ustinoff, pp. 143–75. Newcastle upon Tyne: Cambridge Scholars Publishing, 2012
Vickers, Emma. *Queen and Country: Same-sex Desire in the British Armed Forces, 1939–45*. Manchester: Manchester University Press, 2013
Wadham, Ben. 'Brotherhood: Homosociality, totality and military subjectivity.' *Australian Feminist Studies* 28, no. 76 (2013): 212–35
—— 'Mogan hunts and pig nights: Military masculinities and the making of the arms-corps soldier.' Paper presented at the Australian Sociological Association Conference, La Trobe University, Melbourne, 2004
Wadham, Ben, Donna Bridges, Anuradha Mundkur and James Connor. '"War-fighting and left-wing feminist agendas": Gender and change in the Australian Defence Force.' *Critical Military Studies* 4, no. 3 (2018): 264–80
Ware, Cheryl. *HIV Survivors in Sydney: Memories of the Epidemic*. Basingstoke: Palgrave Macmillan, 2019
Weisse, Major U.J.A. *Report on the Officer Attitude and Opinion Survey: Topical Items 1993–94*. Canberra: 1st Psychological Research Unit, Australian Army, June 1995
Western Australia. 'Report of the Honorary Royal Commission Appointed to Inquire into and Report Upon Matters Relating to Homosexuality.' 18 September 1974
Wharton, James. *Out in the Army: My Life as a Gay Soldier*. London: Biteback Publishing, 2014
Willett, Graham. 'The darkest decade: Homophobia in 1950s Australia.' In *The Forgotten Fifties: Aspects of Australian Society and Culture in the 1950s*, ed. John Murphy and Judith Smart, pp. 120–32. Melbourne: Melbourne University Press, Australian Historical Studies, 1997
—— *Living Out Loud: A History of Gay and Lesbian Activism in Australia*. St Leonards, NSW: Allen & Unwin, 2000
Willett, Graham, Wayne Murdoch, and Daniel Marshall (eds). *Secret Histories of Queer Melbourne*. Reprint edn. Parkville, Vic: Australian Lesbian and Gay Archives, 2017
Willett, Graham, and Yorick Smaal. '"A homosexual institution": Same-sex desire in the army during World War II.' *Australian Army Journal* 10, no. 3 (2013): 23–40
Wilson, Emily. '"Someone who is sick and in need of help": Medical attitudes towards homosexuality in Australia, 1960–79.' In *Homophobia: An Australian History*, ed. Shirleene Robinson, pp. 148–71. Sydney: Federation Press, 2008
Wilson-Buford, Kellie. *Policing Sex and Marriage in the American Military: The Court-Martial and the Construction of Gender and Sexual Deviance, 1950–2000*. Lincoln, NE: University of Nebraska Press, 2018
Wolfenden, John. 'Report of the Committee on Homosexual Offences and Prostitution.' Ed. Scottish Home Department Home Office. London: Her Majesty's Stationery Office, 4 September 1957
Wotherspoon, Garry. *City of the Plain: History of a Gay Subculture*. Sydney: Hale & Iremonger, 1991
—— *Gay Sydney: A History*. Sydney: NewSouth, 2016
Yerke, Adam F., and Valory Mitchell. 'Transgender people in the military: Don't ask? Don't tell? Don't enlist!' *Journal of Homosexuality* 60, no. 2–3 (2013): 436–57

Young, Alison. 'Into the blue: The image written on law.' In *Cultural Analysis, Cultural Studies and the Law: Moving Beyond Legal Realism*, ed. Austin Sarat and Jonathan Simon, pp. 327–52. Durham and New York: Duke University Press, 2003

Young, Rebecca M., and Ilan H. Meyer. 'The trouble with "MSM" and "WSW": Erasure of the sexual-minority person in public health discourse.' *American Journal of Public Health* 95, no. 7 (2005): 1144–9

Young, Warren L. *Minorities and the Military: A Cross-national Study in World Perspective.* Westport, CT: Greenwood Press, 1982

Zeeland, Steven. *Sailors and Sexual Identity: Crossing the Line between 'Straight' and 'Gay' in the US Navy.* New York: Routledge, 2011

4. Media

Mainstream media

AAP General News Wire
ABC (online)
Advertiser (Adelaide)
Advocate (Burnie, Tas)
Age
Age Good Weekend
Argus (Melbourne)
Army
Australian
Canberra Times
CBC (Canada; online)
Central Magazine
Central Queensland Herald (Rockhampton, Qld)
Conversation (online)
Cosmopolitan
Courier-Mail
Daily Telegraph (Sydney)
Guardian
Herald (Melbourne)
Herald Sun (Melbourne)
Independent (Perth)
Mandarin (online)
Maryborough Chronicle (Maryborough, Qld)
Mercury (Hobart)
Mirror (Perth)
Morning Bulletin (Rockhampton, Qld)
Navy Daily
Navy News
News (Adelaide)
News.com.au (online)
RN Breakfast
SBS (online)
Sun (Brisbane)
Sun Herald (Sydney)
Sunday Age
Sunday Telegraph (Sydney)

Sydney Morning Herald
Telegraph (Brisbane)
Truth (Adelaide)
Truth (Melbourne)
Truth (Sydney)
Washington Post
Weekend Australian
West Australian
Woman's Day

LGBTI media

Sourced from Australian Lesbian and Gay Archives, unless otherwise indicated.

At Ease
CAMP Ink (Sydney)
Campaign
Capital Q (Sydney)
City Rhythm (Melbourne)
Feminique (Sydney; courtesy Julie Peters)
Gay Community News (Sydney)
Gay News Network (online)
Klick! (Melbourne)
Melbourne Star Observer
OutRage
PanDA (Canberra)
PinkNews (UK; online)
Pink Thylacines (Tasmania)
Seahorse Newsletter (Victoria; courtesy Julie Peters)
Seahorse Scene (Victoria; courtesy Julie Peters)
Star (Sydney)
Star Observer (national)
Sydney Star Observer
Transceiver (WA; sourced from State Library of Western Australia)

Television

'An Australian soldier's bravest moment.' *Sunday Night*. 7 November 2010. Seven Network

Deep Water: The Real Story. Written by Amanda Blue and Jacob Hickey. Directed by Amanda Blue. Produced by Blackfella Films and Screen Australia. 2016. 87 min

'Discriminating Service.' *Four Corners*. ABC. 3 August 1992

'Lieutenant Colonel Cate McGregor.' *One Plus One*. ABC. 5 July 2013. https://www.abc.net.au/news/programs/one-plus-one/2013-07-05/one-plus-one-lieutenant-colonel-cate-mcgregor/4802564. Retrieved 2 January 2020

Media Watch. ABC. 20 August 2018

Seven Days: Love is Love. 15 February 1966. Channel Seven

Sunday. ABC. 3 July 1994

Index

1st Australian Task Force 28
4th Field Regiment 29
121 Signals Squadron 55

Aboriginal and Torres Strait Islander people, citizenship rights 7
Abrams, Lynn 13
Agostino, Katerina 8
Ahlburg, Ronnie 177
AIDS 88, 151–4
  see also HIV
Akerman, Piers 185
Alamilla Boyd, Nan 12, 14
Allen, Brian 25, 28–9
Altman, Dennis 221–2
Amos, Scott 200
Anderson, Cliff 144
Anderson, George 41
Anderson, Roderic 5
'Anne' 144
Anzac Day 118–19, 223–4
Anzac legend 118, 119
Anzac mythology 8
'Archibald' 101
Archives Act 1983 (Cwlth) 10, 43
Army Health Benefit Scheme 159
Army Soldier Attitude and Opinion Surveys 139
Arrow, Michelle 114
Ashley, David 3
Ashton, Sandy 62, 86, 99
Aulich, Terry 129–30
Australia Hotel, Melbourne 38, 40
Australia Hotel, Sydney 38
Australian (newspaper) 213, 220
Australian Army
  investigation of suspected cases of homosexuality 43
  policy on homosexuality 21, 22, 23, 89–90
  prosecution of homosexuality 21
Australian Council for Lesbian and Gay Rights 129, 145, 146
Australian Defence Force Academy (ADFA)
  LGBTI Staff Guide 219
  LGBTI Student Group 212, 219, 223
  non-binary cadets 206–7, 213
  transphobia 206–7
Australian Defence Force (ADF)
  approaches to LGBTI service 2, 199, 218–21, 233
  culture change since World War II 2, 3, 9
  diversity and inclusion strategy 219–20, 225
  education module to address homophobia 150
  HIV/AIDS policy 88, 151–4, 213–18
  homophobia 125–7, 132, 142, 145–6, 147, 148–50, 158, 201–4
  inquiry into sexual harassment 146–8
  integration of women 9
  leadership combatting homo/bi/trans/interphobia 218–21
  lifting of ban on LGB service 127–32
  members' attitudes towards homosexuality 124–7
  policy on homosexuality 10, 76–8, 82, 87, 90, 113, 114, 116
  policy on transgender personnel 185, 186–91, 208–10
  recognition of same-sex relationships 154–6, 158–9, 162–4, 185
  reforms 9
  sexuality-based prejudice 200–4
  'unacceptable sexual behaviour' instruction 148–9
  see also Australian Army; Royal Australian Air Force (RAAF); Royal Australian Navy (RAN)

Australian Federation of AIDS Organisations (AFAO) 152
Australian Human Rights Commission 2, 189, 190, 207
  see also Human Rights and Equal Opportunity Commission (HREOC)
Australian Logistic Support Base, Vung Tau 30–1
Australian Medical Association 153
Australian national identity 8
Australian Security Intelligence Organisation (ASIO) 63
Australian Transgender Support Association of Queensland 175
Australian Women's Army Service (AWAS) 51
Australian Women's Land Army 50
autobiographies of LGB veterans 4–5

Bailey, Ed 140, 141
Baker, John 159
Bandle, Stanley 41
Barnard, Lance 75
Barnes, F.W. 180
Bartlett, Andrew 162
beats 5, 22, 28, 40–1, 100
Beaumont, Rebekah 155
Beazley, Kim 116
Beck, Kirstin 3
Benjamin, Harry 173
Ben-Shalom, Miriam 112
Berg, Vernon 'Copy' 112
Bérubé, Allan 33–4
'Bill' (RAAF) 99, 141, 142
Bill (Vietnam veteran) 28, 31, 32
Billett, Graeme 201
Bilney, Gordon 130
biographies of LGB veterans 4–5
bisexual service members *see* LGB service personnel
Bishop, Bronwyn 159
blackmail, supposed vulnerability of LGB service members to 64–5, 68, 145
'blanket bashing' 125
Blundell, Alix 85–6, 91
Bobbermien, Nikki 178
Bomford, Janette 51, 55
Bone, Liam 205
Bowtell, Bill 145
Bradford, David 28, 29
Bradford, Riley 211–12, 213
'Brett' 95
Brinkley, T.E. 147
British Armed Forces
  ban on LGB service 156, 234
  diversity and inclusion in the armed services 233
British Army 24, 25
British Royal Navy 34
Brooks, Robert 174
Brown, George R. 170, 175
Brown, James 8
Brown, Rob 201
'Bruce' 93, 99
Bryant, Karl 83, 157–8, 159, 161
bullying
  of LGB service personnel 77, 91, 124, 142, 146, 149, 203
  official investigations of 37
  of transgender service personnel 178, 206
Burns, Austin 41
Butler, Maud 173

C v Commonwealth of Australia 215
Cahill, Craig 100, 101, 132
Cairns, Jim 113
CAMP (Campaign Against Moral Persecution) 110, 111–14, 115
CAMP Ink (newsletter) 67, 75, 111–12
Campaign 116
Campbell, Angus 86, 204
Campbell, Max 118, 119, 223
Canadian Defence Force 131, 146, 233–4
Capezza, Nicole 147
Carlisle, Kate 80–1, 88, 141
Carney, Dennis 61
Carter, Janet 80, 91
Catholicism 26
Caucus Joint Working Group on Homosexual Policy in the Australian Defence Force 129–30, 145, 146
celibacy 26, 28–9
Chameleon Society of Western Australia 173, 174
Chong, Vince 144, 220–1, 222, 223
Cincotta, Loretta 222
cisgender 171
Citizen Military Forces 40
citizenship rights, military service and 7–8
City Rhythm 116, 118
Clark, Darran 90
Clinch, Bridget 188–9, 190, 191
Collyer, David 28, 29, 40
Commonwealth Ombudsman 122
composure *see* oral history

Cone, Matt 125, 142–4, 148
conservative media, attacks on ADF's support for LGBTI inclusion 209, 213, 218–19, 225
see also Murdoch press
Convention on the Rights of Persons with Disabilities 189
Coogan, Eunice 83–4
'Cooper' 202
Cosgrove, Peter 163
Court of Inquiry into Conduct of RAAF Police at Air Force Academy 121–2
courts martial 1, 21, 35, 38, 41–2, 43
Cowin, Wally 33, 35–6
Crews, Bill 163
Croome, Rodney 129, 145, 146, 148, 150
Crump, Elaine 56–7, 62, 80
culture wars 209, 213, 219, 222, 225
Cummings, Katherine 5, 176, 178

Daily Telegraph 219
Daniels, Viv 65
Daughters of Bilitis 111
Davis, Karen D. 233
Davis, Madeline 10, 51
Defence Abuse Response Taskforce (DART) 39, 85–6, 203
Defence Act 1903 (Cwlth) 128
Defence Corporate Directory 219, 225
Defence Force Discipline Act 1982 (Cwlth) 116–17
Defence Force Ombudsman (DFO) 122–3, 159
Defence Force Recruiting (DFR) 192–3, 224
Defence Health 159, 207, 208, 209
Defence Pride Network 223
DEFGLIS (Defence Gay and Lesbian Information Service) 1–2, 161–3, 164, 194, 199, 203, 206, 219, 220–4
Dennis, Christina 69
Department of Defence
reasons for ban on LGB service 53, 122–3
tri-service review of policies on homosexuality 75, 76–7
Department of Naval Intelligence 36
Devine, Miranda 219
D'Hagé, Adrian 145, 148, 149, 152, 160
'Diego' 201–2
Dinnison, Michael 99
Disability Discrimination Act 1992 (Cwlth) 152–3, 189
discrimination 9–10, 113–14, 154–6
dishonourable discharges 22, 35, 36, 42–3, 49, 52–3, 64, 67, 68–70, 93, 96–8
D'Orsay-Lawrence, Marie-Desirée 175
Draft Resisters Union 177
drag performances 173–4
Drew, Keith 83
Duffield, Kyle 80, 84
Duffy, Michael 128, 129, 130
Duniam, Shane 81, 82, 87
Dunlop, John 34, 43
Dunlop, Weary 130

Easterbrook, Leigh 127
Edwards, Colin 91–2, 93–4, 102, 103, 144
Ellicott, R.J. 115
'Elliott' 212–13
Ellis, Kaye Rona 175
Ender, Morten 233
equal rights, military service and 7–8
espionage, supposed links of homosexuality to 63
Ewing, M.J. 120–1

Facebook, pages vilifying homosexual personnel 203–4
Fair Work Act 2009 (Cwlth) 215
Farrugia, Tony 183–4
Farthing, David 90
Fischer, Neal 201, 221
Flanagan, Bob 24–5
Ford, Gen 62, 68, 88, 94
Ford, Ruth 4, 50–1, 52, 61
Four Corners 129
Francis, Arthur 127
Fraser government 114–15
free speech 204
French, Robert 63
Fritzell-Flint, Renae 221
Fry, Leon 30, 86–7, 100–1
Full Spectrum Ltd 142

Gates, R.W. 143
gay and lesbian history, as academic pursuit 5–6
gay and lesbian rights movement 112
Gay Ex-Services Association (GESA) 118–20, 223
Gay Liberation 111
gay service members *see* LGB service personnel
gay/bisexual men
marriage or relationships with women 99
Gaynor, Bernard 204, 222

Gaywaves (radio program) 101
gender identity *see* transgender people
'Geoff' 126
G-Force 156–61, 164
'Gina' 182–3, 184
Gleadhill, Garry 37–9
Goldsby, Tom 24, 25
Goldschmidt, Richard 181
Goward, Pru 163
Gration, Barry 120
Gration, Peter 83, 120, 143
Gration, Richard 83, 84, 90, 101, 120–2
Green, Anna 13
Green, Becky 201
Green, Lyndal 57, 61, 62, 65, 66, 86
Greig, Brian 162
Griggs, Ray 95, 209–10, 218, 219, 220, 234–5
Gulf War 101
Gurkin, Danielle 221

Haas, Connor 201
'Hadrian' (WWII veteran) 5, 24, 25, 32, 44, 112
Hamblin, Amy 188, 189–90, 191, 193
Harding, Donna 191, 192, 194, 208
Hart, Gavin 27
Hastie, Andrew 209
Heal, Catherine 179, 180
'Helen' (Navy) 180
'Helen' (WRAAC) 49, 58
Hendy, Julie 56, 60–1, 64–5, 69, 70
Herbert, Bob 1
Highly Affective Anti-Retroviral Therapy (HAART) 154, 214
Hirschfeld, Magnus 173
HIV 151–4, 213–18
  see also AIDS
HMAS *Australia* 176
HMAS Cerberus 203
HMAS Creswell 37–8, 102
HMAS *Derwent* 126
HMAS *Fremantle* 174–5
HMAS *Melbourne* 33, 36, 37, 92, 102, 180
HMAS *Perth* 38–9
HMAS Platypus 125
HMAS Stirling 126
HMAS *Stuart* 33
HMAS *Supply* 92
HMAS *Swan* 110, 146
HMAS *Sydney* 33
HMAS *Vendetta* 33
HMAS Watson 35, 142
homophobia 116, 118, 126, 132, 141, 145–6, 147, 148–50, 158, 201–4
Homosexual Law Reform Association 111
homosexual service members *see* LGB service personnel
homosexuality
  ADF members' attitudes towards 124–7
  decriminalisation 76, 112, 113, 115
  developmental explanations of 89
  medical model 52, 63–4, 67–8
  policing of 40–4
Honorary Royal Commission to Inquire into Matters Relating to Homosexuality (WA) 32, 43, 112–13
Hosking, Bart 40
Houston, Angus 163
Howard government 155
Human Rights and Equal Opportunity Commission (HREOC) 111, 123–4, 127–8, 152, 153, 159
  see also Australian Human Rights Commission
Humphries, Catherine 188, 191, 194, 205, 221, 235

International Covenant on Civil and Political Rights (ICCPR) 128, 189
International Force East Timor (INTERFET) 162
International Labour Organization (ILO) conventions 128
Intersex Human Rights Australia 172, 181
intersex people
  definition 172, 181
  gender identification 210
  invisibility of 181–2
intersex service personnel
  absence of policy on 170
  experiences 182–4
  gender identification options 210–11
investigations of LGB service personnel
  courts martial 1, 21, 35, 38, 41–2, 43
  dishonourable discharges 22, 35, 36, 42–3, 49, 52–3, 64, 67, 68–70, 93, 96–8
  disproportionate targeting of women 96
  gay and bisexual men 1950s to 1970s 41–4
  initiation of 78–81
  lesbian and bisexual women from 1950s to 1985 52–3
  officers 97–8
  policy on 77–8
  retention in services 90–1

service police interviews 83–6, 120–2
statistics on 95–8
surveillance of suspects 81–3
toll on accused 93–5
voluntary discharges 43–4, 69, 71, 75, 76, 77, 79, 80, 82, 85, 91, 92, 96, 101
witch-hunts 63–8, 86–8
Isaacson, Peter 119
Israeli Defence Force 131

Jackson, Dawn 63–4
Jaggers, A.M. 54
James, Digger 152, 161
Jamieson, Graham 24
Janowitz, Morris 7
Jarmyn, Mike 119
'Jason' 126, 141
Jaynes, Allan Royce 179
Jeffers, Keith 102, 127, 155, 162
Jeffries, Jennifer 61, 67, 71, 95
Jennings, Rebecca 60
Jericho, Jyonah 8
'John M.' 125, 126
Jones, Lynette 174
Jorgensen, Christine 173, 174
judicial activism 112
Junger, Sebastian 28

Kansteiner, Wolf 13
Kardas, Lucy 123–4
'Kate' 178–9
Keating, Paul 129, 130, 145
Keating government 130–1, 152–3
Keenan, Tony 152
Keon-Cohen, Colin 117
'Kevin' 100, 101, 141
Keys, William 117
Killen, James 52, 110, 121
Klinefelter syndrome 5, 182
see also intersex
Korean War 24

Lambie, Jacquie 209
Lapovsy Kennedy, Elizabeth 10–11, 51
'Larissa' 202, 203
Lavarack Barracks 40
Lawrie, Brian 99–100
lesbian service members *see* LGB service personnel; LGBTI service
lesbianism
medical model 52
military services' anxieties about 50–3
Lewis, Alan 28, 40
'Lexa' 202–3

LGB service personnel
affirming experiences 200–1
autobiographies/biographies of veterans 4–5
bullying of 37, 77, 91, 124, 142, 146, 149
celibacy 26, 28–9
discriminatory treatment 9–10, 113–14, 154–6
and homosexual subcultures 33–4, 37–40, 60–2
investigation of. *see* investigations of LGB service personnel
mental health problems 32, 85, 124, 142–3
protection of 37–40, 80–1
retention following warnings 90–1
as security risks 63, 64, 68
sexual assaults 36, 38–40, 79, 126, 142
sexual harassment 79, 123, 146–8, 149
suicides 94–5, 124
tolerance of 9, 22, 23, 24, 25, 27, 29, 37, 44
as vulnerable to blackmail 64–5, 68, 145
LGBTI, term 2
LGBTI service
celebrations of 223–4
percentage of permanent forces 224
'Linda' 200
Liversidge, Danny 81–2, 93
Livingstone, Mark 147–8
Lord, Bill 102
Lummis, Trevor 14
Lupton, Deborah 151

Malayan Emergency 24–5
Manning, Eleanor 51
'Mark' 79, 94, 125
Markussen, Ingrid 149
Marr, David 161
Martin, Stuart 88
masculinity, stereotypes 8–9, 139–40
Mather, Gina 175
Matlovich, Leonard 112
Matthews, Michael 233
McFarlane, Brian 26–7, 29–30, 31–2, 38
McGeorge, John 37
McGregor, Catherine 169–70, 191
McInerney, Sandra 88, 144–5
McKeown, Kenton 92, 125–6
McLeod, Tony 81, 103
Medical Classification Review Board 189, 218

medical model of homosexuality 52, 63–4, 67–8
Melbourne Star Observer 142, 150, 155
memory *see* oral history
mental health problems 85, 124, 142, 189, 207
Mercer, Johnny 234
military cultures
  gendered nature of 8–9
  homosociality 9, 139
Military Pride Ball 199, 223
military sociology 7–10
misogyny 205
'Mitch' 186–7, 206, 207
Mitchell, David 149–50, 156–7, 158, 159, 160, 161
Montesini, Lorenzo 5, 30
Moore, Eugene 142
Morrison, David 26
Morrison, General David (former Chief of Army) 143, 169, 199, 219, 231
Murdoch press 209, 219
  see also conservative media
Murray, Neil 95, 98

'Nathan' 94
National Archives of Australia 10
National Service 50
NATO 128
Naval Board, list of suspected homosexuals 36
Navy Diversity Forum 219
'Nelson' 153–4
New Zealand Defence Force 131
Newman, Jocelyn 159
NSW Gay Rights Lobby 115, 116
Nui Dat base, Vietnam 28

O'Brien, Stuart 1–2, 162, 163, 219
O'Hara, Jim 185
Oldrey, Wayne 152
oral history
  interviews 11–12
  memory and composure 12–13
  theory and methodology 10–14
OutRage 28, 79, 81, 83, 88, 90, 117

Paige, Caroline 3
'Paul' 82
Payne, Marise 210
Peace Monitoring Group in Bougainville 150
Pearson, Cedric Maudsley Ingram 43, 112
'Pencil' 58, 79, 141
'Penelope' 176–7
Pentreath, Mark 222
Percy, Robert 38–9
'Peter' (RAN) 28, 84, 92–3, 222
Pham, Dana 192–3, 194
Phillips, Peter 139
police, targeting of gay or bisexual men 40–1
Pollock, Alex 179–80
Popham, Carole 56, 69
Powell, Janet 128
Prendergast, Gavin 147, 148
PrEP 213, 214–15
Pritchard, Adam 203
Puckapunyal 41, 66

queering of occupations 34
Quigley, Delia 59–60, 61, 64, 65, 69, 101

RAAF Academy, Point Cook 120–1
RAAF Base Edinburgh 87, 102, 127, 183
RAAF Base Glenbrook 79
RAAF Base Pearce 177–8
RAAF Base Penrith 178
RAAF Base Point Cook 81
RAAF Base Wagga 102
RAAF Base Williamtown 145
Radicalesbians 111
Rank Outsiders 156–7
rape *see* sexual assaults
Ray, Robert 128, 130
Returned and Services League (RSL) 117–19, 130, 161, 163
Reynolds, Robert 22
Ridge, Janet 179
Robinson, Trevor 25–6
Rogers, D.N. 124
Rohall, David 233
Roque Ramirz, Horacio N. 12
Royal Air Force (RAF), Exercise Long Look 156
Royal Australian Air Force (RAAF)
  Court of Inquiry into Conduct of RAAF Police at Air Force Academy 121–2
  courts martial 41–2, 43
  dishonourable discharges 42–3
  gay and bisexual men 26, 30–1
  police interviews of LGB service members 83–4, 87, 120
  sexual harassment 79
  surveillance of LGB service members 81–3, 123–4
  transitioning gender handbook 193–4

Royal Australian Electrical and Mechanical Engineers 25
Royal Australian Navy (RAN) 33
  courts martial 35, 43
  dishonourable discharges 22, 35, 36
  Good Working Relationships program 149
  homosexual subcultures 33–4, 37–8
  investigation of suspected cases of homosexuality 34–7, 43
  policies about homosexuality 33, 34, 35, 89
  prosecution of homosexuality 21–2
  sexual assaults 36, 37
Royal Australian Regiment
  1st Battalion 37
  2nd Battalion 40
Royal Commission into Institutional Responses to Child Sexual Abuse 95
Royal Commission on Human Relationships 113–16
Royal Military College, Duntroon 26
Rudd government 155
Ruxton, Bruce 117, 118, 119, 185
'Ryan' 84

same-sex relationships, recognition by ADF 154–6, 158–9, 162–4, 185
'Sandra' 59, 68, 70
saunas 99, 100
scandals in the services 41–2
Seah, Mike 155, 156, 157, 158–9, 161, 162
Seahorse Club 174, 175
Senate Inquiry into Sexuality Discrimination 159, 175, 185, 186
Senate Sexual Harassment Inquiry 145, 146–8
Sex Discrimination Act 1984 (Cwlth) 58, 163, 189, 212, 213
sexism 147, 202–3
sexual assaults 31, 36, 38–40, 79, 126, 142
sexual harassment 79, 123, 146–8, 149
Sexuality Discrimination Bill 185
Sheather, E.H. 147
Shilts, Randy 30, 31
Shrine of Remembrance, Melbourne
  as beat 22
  wreath-laying by GESA 118–19
Signal Corps 55, 79
Sillett, Yvonne 79, 91
'Simon' 90–1, 132
Sinclair, Chris 61, 68, 83, 103
situational homosexuality 31–2, 35, 52
Smaal, Yorick 4–5, 21, 22, 32, 173–4
Smart, Tracy 218, 221
Smith, Hugh 9
social media, vilification of homosexual personnel 203–4
soldier-citizen framework 7–8
Special Investigation Branch (SIB) 66, 79, 80, 81, 83, 85, 86, 102
Spence, Kenny 103
Spencer, Ronni 57, 66, 87, 125
Stankovic, W. 113
Stanley, Peter 21
Steele, B.G. 64
Stirling, Peter 5, 182
Straub, Robert 30
Strawbridge, Nola 56, 58–9, 62, 67
Struth, Susie 59, 66, 70, 86
Stryker, Susan 171
suicides 94–5, 124, 178, 179
Summerfield, Penny 12, 13
Sydney Gay and Lesbian Mardi Gras 102, 127, 159–61, 199, 219, 221–2
Sydney Gay Youth Group 101

Teddy Bear Terry 118, 119
Tehan, Dan 213
terminology 2–3, 171–2
'Terri' 79
Thomson, Alistair 12–13
Timmins, James 41, 44
Tolland, Russ 222
Tovey, Noel 5
transgender people 2–3, 171–2
transgender service personnel
  autobiographies and biographies of 4, 6
  ban on 3, 170, 185–91
  bullying of 178, 206
  challenging of ban 188–91
  enlistment process 192–3
  gender affirmation surgeries 190–1, 206–7, 208–10
  investigations by medical authorities 187–8
  investigations by service police 177–8
  medical discharges 180–1, 183, 184, 189
  motivations for enlisting 170, 175–6
  in National Service 176–7
  peer support and information-sharing 194
  re-enlistment following transition 186–7
  struggles with gender identity 3, 170, 188
  transphobia 204–8
Trembath, Richard 55

'Trent' 215–18
Trudeau, Justin 233–4
Trump, Donald 225
Truth newspaper 21, 41–2, 87, 91
Truvada *see* PrEP

United Kingdom
  government apology for treatment of LGBT veterans 234
  see also British Armed Forces
United Nations Human Rights Council (UNHRC) 155
United States
  'don't ask, don't tell' policy 131
  military activism 112
  transgender military personnel 225
US Women's Army Corps (WAC) 51
Ustinoff, Julie 58

van Netten, Anna 66–7, 71
'Vicki' 179
Victoria Barracks 38, 102
Vietnam War, homosexual relations during 27–32
voluntary discharges 43–4, 69, 71, 75, 76, 77, 79, 80, 82, 85, 91, 92, 96, 101
von Clausewitz, Carl 7
Vung Tau, Vietnam 27–8, 30

Wadham, Ben 8–9, 139–40
Walls, Robert 160
Watson, Alexander 'Lex' 115–16
Watson, Stuie 99, 157
Wells, David 140
Whelan, Tony 26
Whitlam government 75, 113
Willett, Graham 22
Williams, Richard 37, 94
Williams, Susan 174
Wilson, Joel 194, 205, 206–7
Wilson-Buford, Kellie 4, 8
witch-hunts 61, 63–8, 86–8
women
  integration into ADF 9
  role and status in 1960s 58
  see also women's services
Women Against Rape in War 118–19
Women's Auxiliary Australian Air Force (WAAAF) 54
women's movement 58
Women's Royal Australian Air Force (WRAAF) 50, 54
  dishonourable discharges 52–3
  promotional material 55
Women's Royal Australian Army Corps (WRAAC) 80
  dishonourable discharges 49, 64, 69
  first overseas deployment 55, 69
  operation 50
  policy on homosexuality 56, 63–4
  promotional material 54–5
  total enlistments 54
Women's Royal Australian Naval Service (WRANS) 50, 54
  dishonourable discharges 68, 69
  policy on homosexuality 56–7
  promotional material 54
  voluntary discharges 80
women's services
  dishonourable discharges 52–3, 67, 68–70
  investigation of suspected cases of homosexuality 49–50
  lesbian subculture 49, 60–2
  motivations for joining 49, 51–2, 58–60, 64
  official anxiety about lesbianism 50–3
  scope of work 55–6
  sports teams 62
  tactical significance 50
  unacceptability of homosexuality 56–7
  witch-hunts 61, 63–8, 88
  working conditions compared to men's services 55
  during World War II 50
Woodside Barracks 40
Woollard, Keith 153
World Professional Association for Transgender Health 208–9, 210
Wotherspoon, Garry 21, 22

X v Commonwealth of Australia 153, 213
'Xavier' 33–4

YouTube 203–4
Young, Alison 153
Young, Edward 155